THE DETROIT RIVER

Ontario and the Detroit Frontier
1701-1814

Ontario and the Detroit Frontier 1701-1814

by

Hugh Cowan M.A., B.D.

(Originally published as "Canadian Achievement in the Province of Ontario, Volume 1: The Detroit River District")

****

~PLEASE NOTE~

This book was first published in 1929 under the title *Canadian Achievement in
the Province of Ontario Volume 1: The Detroit River District*, by Hugh
Cowan, M.A., B. D. The title has been changed on this reprint edition to better
reflect the contents for a modern audience. Upper Canada Press has not edited
or altered the text, and can assume no responsibility for the accuracy of the
information contained herein.

****

Upper Canada Press

An imprint of

American History Press

Franklin, Tennessee

(888) 521-1789

Visit us on the Internet

www.Americanhistorypress.com

ISBN 13: 978-0-9830827-4-3

Library of Congress Control Number: 2012935813

Printed in the United States of America on acid-free paper.

This book meets all ANSI standards for archival quality.

Introduction

The decision to reprint this book was based on the quality of its contents and the time period that it encompasses, as well as the 2012 celebration of the bicentennial of the War of 1812 in Essex County. When researching county history, local historians have primarily relied on three books: *The Windsor Border Region* by Earnest J. Lajeunesse (1960), *Uppermost Canada* by R. Alan Douglas (2001), and *Garden Gateway*, by Neil F. Morrison (1954). These three works cover our region's history from 1701 to 1950, and are the most informative books available devoted to that time period. *Ontario and the Detroit Frontier 1701-1814* embraces the entire history of the region from 1701 until the end of the War of 1812. As such, it will be a welcome resource for all those who enjoy local history, and will help consolidate the information in the other three books into one volume.

The main interest of this book for all readers is the fact that all of the major events that affected Essex County and the "Northwest Frontier" are discussed in a readable and easy to understand style. Essex County was unique in the sense that it went through a variety of name changes and was controlled by three different countries with distinct political systems, religions and beliefs. In *Ontario and the Detroit Frontier 1701-1814* you will read about the political and social history of the Detroit Region and the Western District, and learn of the principal people involved in the development of this section of modern Ontario.

Among the multitude of topics covered is Champlain's discovery of the area in 1701, the reign of the French until 1785, and the British/French war that culminated on the Plains of Abraham in a major confrontation. After the solid establishment of British rule in 1759, the author explains the key influences of the American

Introduction

Revolution, followed by the exodus of the British settlers from the state of Michigan. He also describes in detail the "New Settlement" in 1796 along the north shore of Lake Erie, followed by the construction of Fort Amherstburg (Malden), and the participation of local Indian tribes in the entire process. He then explores the complexities of the War of 1812 in an area that was under American rule for almost two years, with the "state" capitol being Sandwich Towne and the Duff/Baby house the "White House," with the future American President William Henry Harrison in charge of affairs there.

According to the author, Sandwich Towne is a "place of historical interest," and many of today's historians will agree with this statement. In this narrative you will discover the hidden history of our area and, hopefully take the time to explore the places you learn about, thereby giving you more appreciation for our unique and wonderful country.

Chris Carter
Upper Canada Press
Colchester, Ontario
April 2012

CANADIAN ACHIEVEMENT

"NO good history has been, or can be, produced in haste." "The collection of materials, if nothing to perfect it be left undone, will require much time, trouble and expense."

These two quotations describe succinctly the task which the Algonquin Historical Society of Canada has undertaken to accomplish—the writing of the story of the progress and development of Canada, beginning with the province of Ontario, under the title, "Canadian Achievement."

This is a great and difficult task, because of the extent of the field which it is intended to cover, and the accuracy, even to the minutest details, which must be obtained as to facts, if it is to be made a valuable and serviceable undertaking. The task, though great, is not impossible, if energetic and persistent voluntary leadership be given and generous contributions of money made, by those interested in preserving the country's history.

The method is to trace the development of the Province, section by section, following in the main, the geographical and political communities into which the Province has been subdivided.

This volume is the first of the contemplated series. It narrates the settlement and development of the Detroit River, or Western District, until after the close of the War of American Invasion, 1812-1815, the locality where, in 1701, was established the first French colony of the west, the story of which, and the subsequent developments therefrom, constitute the first chapter in the history of the Province.

Dedicated to the Pioneers and their
Descendants who laid the Foundation
of a New Nation in this section of
the Dominion of Canada and
Province of Ontario

CONTENTS

arouses Indian Hostility—American Indian Warfare affected Canadian interests—The leadership of the Fur-traders—Two classes of Traders at Detroit—Neutrality of Fur-trader Impossible—Influence of These Traders Under-rated—Dissatisfaction with Treaty of 1783—Canada Unrepresented on Commission—The Fur-traders Resent Their Findings—The forfeiting of British property rights also resented—The Treaty stands eleven years in abeyance.

CHAPTER IX.

The Detroit River District is named Hesse—The first Court of Justice within its bounds, established by Proclamation of Lord Dorchester, 1788—Districts formed for Civic Purposes —The first Court Officials—William Dummer Powell becomes first Judge of Ontario—Opens a law-office in Montreal, 1779—Agitates for Revision of Quebec Act—Friendship of Lord Dorchester—Difficulties to Maintain Strict Integrity in the West—Withdraws from the troubled political atmosphere of North America—Plots to bring about his ruination—Author of Brock's Proclamation — Subsequent Offices and Emoluments—Domestic Anxieties and Troubles —Death and Burial, Aged 79.

CHAPER X.

The Pioneer Effort to Establish Municipal Government in Upper Canada, by the Organization of Districts and Counties—The Boundaries of the Province Undetermined—The Establishment of Counties—County Lieutenants Appointed for Civil and Military Purposes—The First Boundaries of the Western District—The Second Boundaries of the Western District—The Third Boundaries of the Western District— District of Kent Formed, 1847—Districts Abolished, 1849— The Electorate of Bothwell Organized, 1877.

CHAPTER XI.

The Jay Treaty and the Exodus of 1796 — Circumstances Leading to renewal of War—Governor Simcoe Would Stand Aside Neutral—American Conflict with the Indians—General Wayne's Activities on the Miami—The Wanton Destruction of Indian Agent McKee's Property—Canadian Militiamen Assist the Indians—Governor Simcoe's Military Prescience—President Washington Smooths the Way for Peace —Chief Justice John Jay, Plenipotentiary—Reception Accorded His Successful Undertaking.

10

FOREWORD

***

The Story of the Canadian People and Their Achievements in the Detroit River District

OF the making of books, there is no end. Nor indeed can there be. As long as there is progress, something new added to the world's history, so long will there be a need to put this on record. History is but the taking stock of ourselves, what we were, what we are, and as a deduction from these, what we may expect to become. If it is true, that those who would make history, must first study history, then we must record progress and achievement, in order that further progress and achievement may be attained.

A nation's achievements are but the sum of the achievements of the individual units which comprise the nation. From the viewpoint of some, the country is achieving progress only in the measure in which it is accumulating material wealth. With these, a nation is not following up a progressive aim unless it is building great cities, great financial corporations, great industries, and great fortunes for a selected few. From the viewpoint of others, the test of a country's greatness is in its military power, the efficiency and strength of its navy and army, as France in the days of Napoleon, Rome in the days of Julius Caesar, and Germany under the tuition of Bismarck.

History corrects the thinking of these and bids man find true progress in the development of man's own personal life. 'The proper study of mankind is man.'

Hence our treatment of the story of Canadian progress, must of necessity be from a biographical viewpoint, discovering the men who comprised the successive and separate links in that long chain of events which contributed their part to the development and progress of the whole.

* * *

Introductory Analysis of the History of the District until the Close of the War of American Invasion

THE first two centuries of Canadian history is a story of constantly recurring wars. It was a struggle for the posession of the land—who should own it, and to what race of people it should belong?

At first the struggle was between the French and the Indians; then between the French and the British, the Indians taking the side of the French. After getting possession of the whole of North America, the British became a divided house. British fought against British on two successive occasions. In the second instance, the British domiciled in the United States fought against the British domiciled in Canada, the Indians taking the side of the Canadians. Thus, from the landing of Champlain in Upper Canada in 1615 until the close of the War of the American Invasion of Canada in 1815, the country had no rest, but was either actively engaged in war, or in constant expectancy of it.

The first Europeans to get a foothold in the western peninsula of Upper Canada were the French. In 1701, Cadillac, the commander of the French fur-trading post at Michilimackinac, established a second western post on the site where now stands the city of Detroit. Soon this became the centre of Indian trade and Indian settlement. Four tribes of Indians soon domiciled themselves in the immediate neighbourhood of the post—Hurons, Ottawas, Ojibways and Pottawatamies.

Contrary to usual French procedure, a colony of husbandmen was here, also, established. About 1750, this colony extended itself to the Canadian side of the river, and save where the Huron Indians had a reservation at Montreal Point, this settlement by 1796, extended all the way from Lake St. Clair to the Canard river. These comprised the first agricultural settlement established within the bounds of the present province of Ontario.

The Detroit post, with its neighbouring settlements, were held in possession of the French for about sixty years. The fall of Quebec effected by General Wolfe, and the consequent surrender of all the French possessions in North America to the British, brought about a change of ownership in 1760. In the meantime, the fur-post with its accompanying settlements, comprised a colony of twelve hundred in population, of whom three hundred were on the Canadian side of the river.

In the first three years of British possession, the garrison which manned the fort was in a constant state of danger. The Indians were restless and the French of indifferent loyalty. A conspiracy to drive out the British under the leadership of a tribe of the Senecas, was nipped in the bud. A second one, under the leadership of Pontiac, an Ottawa Indian chief, was more successful.

Supported by a confederacy of Indians, he undertook to massacre every British garrison in the West, and was successful every where but at Detroit. After being foiled in two attempts by Major Gladwin, the commander in charge, he put the Fort under a state of siege. This he kept up for one hundred and fifty three days, the longest period in continuance which any Indian endeavour is known to have persisted in order to accomplish its hostile aim.

That summer, the treaty of peace between the French and the British was finally ratified. A copy of its terms was sent to Detroit, read to an assembly of the French, who immediately exhibited a changed attitude to the British. The Indians, then realizing that they could no longer count on French aid, withdrew, one tribe after another, and Pontiac was finally compelled to raise the siege.

For twenty years, the British held possession of Detroit and surrounding territory on both sides of the river, but in 1779, under the leadership of George Washington of Virginia, and others, there was initiated a revolutionary war to overthrow British government in North America, and establish a Republican form of government in its place. This war, from the first, took upon itself the nature of a civil war. An agreement of peace was arranged in 1783, and the Republican form of government acknowledged.

When the treaty of peace was under consideration, Dr. Benjamin Franklin and Mr. John Jay, America's representatives, demanded the cession of the whole of Canada, including Nova Scotia, to be incorporated into and become a part of the newly-created Republic. To this, Mr. Richard Oswald, as representing the British government—the Canadians had no representation—agreed. This was, of course, refused by Great Britain, but when the terms were modified to include the North West, only, all that triangular region north of the Ohio river, out of which there has been carved since, five of the largest states of the American Union, this was accepted as a generous basis of agreement.

But matters did not work out as smoothly on this side of the ocean as anticipated. Resentment was created by the treatment accorded to the Loyalists, that part of the American people opposed to Republicanism, as not in accord with the terms of the treaty of peace. The treaty therefore, continued for some years to be a treaty on paper only. The garrison posts on the frontier, formerly Canadian, now ceded to the United States, continued to be manned by British soldiers, and matters continued to be in this state of unsettlement for eleven years.

In 1794, Mr. John Jay was sent as an American emissary to Great Britain, to discuss terms of agreement. The Grenville ministry, friendly with the United States, agreed among other things, to the cession of the North West and the evacuation of the frontier posts, while the Americans on their part agreed to the payment of debts due to refugee Loyalists.

In preparation for the handing over of Detroit to the Americans, Sandwich was founded as a seat of Government for the Western District, while Amherstburg, because of its strategic position, commanding the entrance to the Detroit river, was selected as the place where the military post and naval station would be established. Some delay in the erection of the fort there occurred, by reason of the United States laying claim to Bois Blanc island, where it was originally intended that the fort should be built. In the meantime, while awaiting decision, the fort was commenced on the mainland, and was there situated when the war of the American Invasion took place.

The first British settlement established in the Western peninsula was made up chiefly of disbanded British soldiers of the Butler's Rangers. They were given holdings on the Lake Erie shore east of Amherstburg. Others of them were settled on the Thames river. These and their sons supplied many of the officers for the militia regiments raised during the war, 1812-1815.

The office of County Lieutenant, created by Governor Simcoe, to keep an enrolment of the men available for military duty, and to call them out for inspection and drill once a year, was still in vogue in 1812. These settlements, together with residents of Sandwich and Amherstburg, were organized into three corps, the Kent regiment, the First Essex regiment, and the Second Essex regiment. The full strength of these units would not be more than two hundred and fifty men each.

When the evacuation of Detroit took place, 1796, few, if any, of the French changed their places of residence. On both sides, the French habitant accepted the rule under which the changing circumstance of the times happened to place him. Many of the British also remained in Detroit, and became subjects of the American Republic.

The policy of Governor Simcoe to have a fleet of twelve gun-boats for Lake Erie, the Detroit river and the northern lakes was not carried out, but his creation of the office of a Commodore was continued, and under him were kept a few gun-boats and a merchant marine available for defensive purposes.

Such, in brief, is the history of the progress and the status obtained by Upper Canada West, when the United States undertook its invasion in the month of June, 1812.

LIST OF ILLUSTRATIONS

THE PIONEER MOTHER

Establishing the foundations of Canadian Progress
on the basis of Industrial Thrift

CHAPTER I.

The Genesis of Canada

CANADA stands for the "happy medium" between the extremes of two political faiths. The seventeenth century witnessed a struggle in Great Britain between the two political parties representing these two faiths. The one party contended that the country should be ruled only by leaders, who, by virtue of their royal birth had the right to rule. This party had the weight of tradition and age-long custom to support it in its contention. There was another party, and at that time a new-born party, who preached the doctrine of self-government—that the right to rule the people must be given to the rulers, by the people themselves, and that they must exercise this right in trust for the people and for the people's common good. The one emphasized the rights of the many to have their desires and will recognized; the other emphasized an inherited leadership of a few, ruling sometimes according to the dictates of their own caprice, and with out any regard to the will or desire of the many. The government of Great Britain came to represent the happy medium between the extremes of these two political faiths.

The "Happy Medium," the truest path to follow

This progressive movement, or awakening era, divided the citizenship of Great Britain into two distinct and opposing groups, each providing a goodly representation of extremists.

On the one side, there was the hardheaded and immovable traditionalist; on the other, the iconoclast. Both of these were defenders of truth and wisdom, but only that which was one-sided.

The traditionalist ignored the rights of common men to have their will and desire recognized in national affairs, stressing only the doctrine of inherited leadership. To some degree, the traditionalist was right. The man capable of leadership must have been born that way. Without an endowment of the gifts suited to the task, no amount of training or self-assertion could supply the qualifications necessary to make a man a good ruler or leader. "It takes the seventh son of the seventh son to make a good glass-blower." In the development of man along a certain line, generation after generation, a proficiency is obtained along that line far above the average.

The doctrine of inherited capacity for leadership may have been overdone as far as our kings are concerned, but he who ignores the necessity of a possession of inherited gifts for successful achievement along any line, is shutting out from his view a most essential requirement, which Nature and history have combined to prove indispensable. If poets are born, not made, so also we have kings who were kings because they were born that way. The traditionalist clung tenaciously to this law and refused to see any exceptions to it; and this faith, was the conviction of more people than the Scottish House of Stuarts.

The iconoclast, seeing a measure of tyranny in the rule of some of the reigning monarchs, came to the hasty and false conclusion that monarchism, in itself, was an evil. When he saw the institutions of the country, both educational and religious, established for the benefit of the privileged class only, he deemed that the cause and betterment of the common people could be brought about only by the destruction of their institutions. Thus on the opposite pole from extreme traditionalism, there grew up out of the revived English life of the 17th century, a radicalism that was rabid, extreme and irrational. The pendulum of the clock swung to the limit of its orbit on the other side. This radicalism found a congenial home in the colonial atmosphere of North America.

Aristocracy and Traditionalism, of 17th Century attacked

The political faith and prejudice of both of these groups, the emigrants took to North America with them. The high ideals which later historians have attributed to the revolting colonists makes good and pleasant reading for their descendants, but the liberty which they are supposed to have fought for, and obtained by the power of the sword, this was a gift handed down to them, by welfare workers from Great Britain. The Seventeenth Century saw the rise of many reformers in the old country, toiling first in the field of religion, and logically next, in the field of politics, to better the conditions of their fellowmen, intellectually, morally, spiritually, economically and socially. The efforts of these to lift their fellow countrymen upward was not permitted to go on, unopposed. The privileged class looked askance at the awakened intelligence and rising aspirations of the common people, the response given to the preaching of these welfare workers, and the changed attitude of mind to monarchical institutions by the lower strata of British society.

The traditions of the church were the first to feel the impact of this new awakening. Ecclesiasticism, however, did not have long to wait for defenders. Opposing men rose up from amongst both clergy and laity, who were prepared to go to the length of burning these welfare-workers at the stake, this, as the easiest way of answering their logic, and curbing their efforts towards self-assertion. The friends of progress, however, notwithstanding the piles of blazing fagots awaiting them, preached on, urging the masses forward to self-assertion and independence, and bade them look to the Bible and its Author, and not to an aristocratic church, for their inspiration and hope.

Who Should be the Rulers and Who the Ruled.

Not content with their attack on the traditionalism of the church, these reformers attacked next the political traditions of their age. Men were not born, they said, some to rule and others to be ruled. The right to rule was not handed down to a special class by Heaven. This was a trust, which had to be acquired by merit, and bestowed upon the qualified by the will and desire of the people. It was not an heritage endowed by birth to successive generations of men. Neither royal blood nor land-ownership counted for anything if the twin qualifications of intelligence and character were wanting. These two conditions supplied, the people themselves, and the people only, had the right to choose their rulers, and instruct them as to the manner and method of their government.

Revolution in America

The opening up of America for colonization, provided a place of refuge for those who had earned the odium of the Monarchical party. In this new land, freed from the restraints of custom, tradition, and of law also, they were able to develop their new faith, unhindered and unopposed. The rapid growth of these non-conforming cults in Great Britain, showed how ready the people were to receive this new teaching. New disciples and converts were daily added, and that all the more easily when they realised, if they were not permitted to put their faith into practice in the Old Land, they could in the New. In this way, America came to receive the best representatives of the new faith which Great Britain provided. These carried with them not a little of resentment at the treatment meted out to them by the privileged classes of Great Britain. It is one of the weaknesses of the human race that they love a martyr's crown, whether or not they have earned it. The opposition of the privileged class, and the sufferings endured by the progressive elements of the community, were re-hearsed to their children with such exaggerations as to lead them to believe that Great Britain was the home of tyranny, not the mother of a progressive, virile race; rising up to a new and glorious era in its history, greater by far than anything hitherto achieved. The pangs attendant upon re-birth were magnified, but the nobler race, arising out of that re-birth, was treated indifferently or ignored entirely. In this way, the colonists in America obtained a ruling majority who represented the extremists on the side of progressive change.

War Declared by the Revolting Colonists, 1779.

This awakened spirit and re-birth of British Society took place decades before the first successful English colony was planted in America. The interference of Old Country leaders in the affairs of the colonists, was practically a negligible quantity. In all matters of vital importance they were free to determine their own destiny. They were an independent nation in everything but the name. Having the reality, they

went to war to obtain the name also. They did not need now
the sheltering care of the Motherland. With the French
beaten and driven out of the country, there was nothing to
hinder them from demanding the full status of nationhood.

Why the Demand for Independence was not Readily Granted

There should not have been any refusal on the part of
Old Country leadership, to have granted them this recognition
without resort to war. But that which gave strength to the
War-party in America, gave strength also to the War-party
of the Old Country. The Reform party was being yearly weakened by the migration from Great Britain to America, for the
major portion of the migrants were from the Reform section
of the Old Country people. The Tory element was in that
measure given greater influence. For this reason, the
counsels of wisdom, which would have prevailed, had there
been a fuller measure of consideration given to American
demands, were ignored. The war-parties in both countries
took matters in their own hands, and their differences had to
be settled by the arbitrament of the sword. Had moderate
men, the determining influence in America, they would have
obtained full status of nationhood without resort to war.

The Course of the Awakening in England.

But while the seed of religious and political freedom,
germinated in Great Britain, grew rapidly when transplanted
to the soil of America, the new faith did not depart from Great
Britain with the embarkation from Plymouth of the Mayflower
with its shipload of refugees. The roots of a tree are strongest
on that side against which the wind blows. Opposition to the

tenets of the new faith and to the new aspirations of the British people, only gave to its proponents greater courage and more determined purpose. That which had been begun in England continued to grow and make its influence felt in every activity and walk of life. Greater than the discovery of America by the British people, was the discovery of themselves—the discovery of their rights and powers. These rights and powers they were determined should be recognized. And recognized they came to be, not only in the setting in order of their own household, but in their relationships with other nations, especially those planting colonies in America.

Destiny of Canada, shaped by Treaty of Peace, 1763

The recognition of their growing strength, the British wrested from the Spaniards, when they defeated them in the war commenced in 1739. With greater difficulty and a longer war, but with the same results, they fought against the French. It was the defeat of this nation and the loss of their possessions in North America, which brought about the organization and establishment of Canada as a British colony.

Canada had a history of its own preceding the year 1763, but it was as a French, not as a British dominion. In that period, the foundation strata of its nationhood had been partially laid. When the French planted their first and successful colony at Quebec in 1608, they began what was destined to become a new nation on the northern half of this western continent. The success which attended that undertaking made the French a permanent part of this new nation. Although their colonizing effort came to an end with the fall of Quebec in 1759, yet by this time they had succeeded in establishing a number of permanent settlements in the country, which they were compelled to hand over to the British by the misfortunes of war. The descendants of these settlers still occupy the land. Others have come from France subsequently to swell their numbers. Their ideas and ideals still influence the character of our national life. In one of our provinces their religion and language is maintained intact, a privilege granted them by treaty, and recognized by the terms of our Constitution in the Confederation Act of 1867.

Tolerance is not a Matter of Choice, but of Necessity

In this way there have been established conditions in Canada which are found neither in Great Britain nor in the United States. We have problems to solve and duties to perform, from which both of these nations are free. The control of our national affairs must be, in consequence, in the hands of moderate men. To put our destiny in the hands of extremists, such as the Revolutionary War exemplified, would be, not only to destroy the peaceful and quiet progress which we are now enjoying, but it would also make shipwreck of all of our future hope and aspirations. These conditions of national life, brought to us from the historic past, compel us to choose, in all our national affairs, a spirit of compromise and tolerance. The second strata of Canadian nationhood came from the United States of America, and as a direct result of the Revolutionary War. The United Empire Loyalists represented neither extreme, but the "happy medium." They asked for self-government, full and free, but this was to be exercised as an integral part of the British Empire, not by the complete severance of the bond between them. When these suffered defeat at the hands of the Revolutionists, a contingent came over to Canada, where they formed settlements, began the country's development, and the development of its institutions in harmony with this policy. Their political faith and policy has been Canada's ever since, accepted by succeeding settlers, not because it was preached and practised by the Loyalists, but because in their own judgment, it was the wisest and best political course to follow.

Early Settlers Evince a Strong British Sentiment

The Britishers who gravitated to this country from the old in the early years of its history, were those who felt strongly in respect to the ties that bound them to Great Britain, whether they were English, Welsh, Irish or Scotch. In many respects, conditions for settlement were much more attractive in the United States than in Canada. The settlers, who came to Canada, chose, what appeared then, a more unfavorable opportunity and this because of the loyalty which forbade them to choose anything other than their own flag. Although developing a new nation on a new continent, Canada's associations with the Motherland have been maintained intact; and thus her associations with the hoary past remain unsevered. To the Canadians of to-day, the handwriting on the wall points out for their country, a career, in which all the wisdom of both political faiths will be found featured, but the irrationalism and folly of their extremes, happily lacking. No change from that policy, visible on our political horizon, it may safely be presumed that on the basis of the "happy medium" our future destiny will be realized. Since the Great World War, a national consciousness has been increasingly to the forefront, due to the important and successful part which Canada and Canadian soldiers took in that struggle; due also to the inspiring prospects for future greatness which a reasonable faith visions as awaiting the development of the country's resources. The present century will doubtless see the enterprising hope of that vision transformed into an accomplished reality.

A question which naturally suggests itself is, whether Canada shall continue to maintain itself a nation predominantly British in the origin, sentiments and spirit of its people. There has been a steady increase in the population of the country from its genesis. This growth in numbers has been brought about by natural increase in a measure, but chiefly by the arrival of immigrants mainly from the British Isles. The first census taken after Confederation revealed a population of 3,689,000. The population of 1928 is estimated at 9,658,000.

For some time, the ratio of British immigrants to those of other nations has been steadily on the decline, until now the numbers arriving annually comprise only 37 per cent of the total. During the few years immediately before the Great World War, there was a rapid increase in the number of immigrants coming into Canada, but during the war and since, there has been a steady decline in their numbers. In the year before the War, 150,542 immigrants arrived from the United Kingdom; 139,000 from the United States; and 112,880 from other countries. During the eleven-year period, 1915-1925, the average number arriving annually were 36,982 from the United Kingdom; 41,412 from the United States; and 21,069 from other countries. In accord with these figures, in respect to the movements of population from other cuntries into Canada, the place hitherto held by the British Isles is now occupied by the United States. But these figures may not be a correct gauge, seeing that these years can hardly be deemed normal, due to the war and the unsettled condition which it produced throughout the world.

CHAPTER II.

The Birth of Ontario

T HE history of North America commences with the year 1492. This is exactly three centuries before the birth of Ontario. As a distinct and separate province, Ontario came into existence in the year 1792. Provision was made by an enactment of the British Parliament, 1791, called the Constitutional Act, for the separation of Upper from Lower Canada, and the establishment of a government of its own, separate and different from that of Lower Canada. The next year, 1792, saw the provisions of this Act brought into effect, and the coming of Col. J. G. Simcoe to Upper Canada, as the province's first Governor. The first parliament of Ontario convened at Newark in the autumn of that year. 1792 therefore, is to us, the beginning year of our provincial history.

Although this date supplies the definite period of time in which Ontario became an integral and separate political unit, organized as an independent province, we cannot, however, plunge into a narrative of its story dating its first historic event from that year. There were events which occurred preceding that period leading to the establishment of Canada as a separate British Colony north of the Great Lakes, and thus laying the foundations for the eventual organization of Upper Canada into a separate province. These should be reviewed if the reader would understand the causes which led to the passing of the Constitutional Act of 1791, and the establishment of Ontario as a distinct and separate province from Quebec, with different laws and government.

The Accidental Discovery of a New Continent

When Columbus set sail from the port of Palos in Spain, in 1492, his aim was not to discover a new continent, but to find a route by means of which Europeans could carry on trade with East India. Before this, a splendid trade had been established, controlled chiefly by the two Italian cities of Venice and Genoa. The route of trade established by the merchants of these two cities was partly by water and partly overland. The caravans of these merchants, the Turks began to attack and pillage as they passed along their overland route, and thus made it impossible for them to carry on their trade unless a new route of travel was discovered. The immediate object of the voyage of Columbus was the discovery of such a route. In seeking it he unexpectedly discovered a continent, the existence of which was hitherto unknown to Europe.

It was not long until the leading nations of Europe ascertained that this continent was very sparsely inhabited; that its resources provided unlimited opportunities for trade; and that it had a land area which was capable of providing homes and food sufficient for their over-populated millions.

Western Europe takes possession of America

Enterprising activities in the realm of seamanship were set on foot by the leading nations of Europe as a result of this discovery. Attempts were made by five separate countries, with more or less of success, to plant colonies in this new-discovered and practically unoccupied continent.

Among these nations, in the first stages of its history,

France occupied the foremost place. French adventurers explored the country and added new territory every new year to the map of North America. French voyageurs went far afield and established an immense and lucrative fur-trade with the Indians. French soldiers built outposts and fortifications to defend their fellow countrymen from attacks by unfriendly Indians. From the shores of the North Atlantic to the headwaters of the Mississippi, and south to the Gulf of Mexico, these outposts, the centres of their trade and commerce, were established. By the middle of the 18th century, the greater part of the continent was claimed by the French as theirs by right of discovery and occupation.

Next to the French came the Spaniards. These settled in the south, first, and from there worked their way northward. They claimed Florida, and disputed possession of the land west of the Mississippi with the French.

In the meantime, British, Dutch and Scandinavians had planted small colonies of their countrymen along the shores of the mid-Atlantic. The British were the last to get a foothold. Their first attempts at colonization were failures. Notwithstanding, they persevered and gradually grew in numbers and strength, until, finally, they became the dominant race in North America, eventually driving out all their competitors.

Rival Nations fight for possession

In the first years of the continent's colonial history, there was no strife or warfare among the representatives of these five nations. The country was large; the colonists were few in number, and set so far apart that they did not in any way interfere with one another. These conditions, however, soon changed. When once a successful British colony had been established, immediately after, Britishers from England, Ireland, Scotland and Wales, began to pour into the country, and to so extend themselves as to soon encroach on the territory of their neighbours. Disputes were settled with a flint-lock musket. The nation which could not win on the battle-field could not permanently hold North American territory. During the whole of the 18th century the colonizing nations in North America were fighting each other, both in the Old Country and the New.

France and Great Britain, 1753-1760

In 1739 war broke out between Spain and Great Britain, ending in victory for the British. In 1755, the British colonists opened out war against their French neighbours of the Ohio River district. Some years before this, the British had pushed their frontier as far northwest as the Ohio river. A Land Company was given a grant of five hundred thousand acres on the south side of the Ohio, on the condition that one hundred families should be settled thereon within seven years, and a Fort erected for their protection. The inspiring mind behind this movement was Governor Dinwiddie of Virginia. The French retaliated by building a fort on the French river near Lake Erie, and another at the confluence of the Alleghany and Monongohela rivers, where they join to form the Ohio river. This post they strongly garrisoned and named it Fort Duquesne. The claims of the French to this territory were inscribed on leaden plates, and deposited at various places throughout the country.

In 1753, Governor Dinwiddie sent George Washington as his messenger to the French Commandant at this Fort, to claim for the British all the territory of this district and warned him, in particular, that he deemed the erection of that Fort there, an encroachment on British rights. An occasion for the continuance of these hostilities was provided by the breaking out in the Old Country, of the Seven Years' War between Great Britain and France. The expedition of 1755, an army in the main of British colonists, with General Braddock in command, against the combined strength of French and Indians at Fort Duquesne, ended in a complete defeat for the British. General Braddock was slain, and his whole force routed.

But the British were not easily discouraged. In the person of William Pitt, they had an energetic leader, who was determined to bring about the defeat of the enemy. The conduct of the war in America he placed under the leadership of Generals Amherst and Wolfe. An adequate supply of men and of equipment was provided them. As a result, fort after fort was compelled to surrender; Forts Frontenac and Duquesne in 1758; Niagara, Ticonderoga and Quebec in 1759; Montreal in 1760. The French were utterly defeated. The Treaty of peace was not made until three years later, 1763, at which time France ceded to Great Britain all her possessions in North America, save New Orleans, and a small district adjacent to that city. Military rule was established in Canada, but the French people inhabiting it were treated with magnanimous generosity by those now in authority over them.

Quebec Act, a Generous Concession to Canadian French

In 1774 the Quebec Act was passed by the British Parliament. By its provisions, the laws of Old France were made the laws of Canada; the French language was permitted to be an official language of the country; the Roman Catholic

religion was sanctioned and the authority of the Papal See at
Rome recognized. Canada, by this Act, was permitted to
remain a French province in everything but name. These
provisions of the Quebec Act found no favor with the British
Colonists in America, especially those of the New England
States.

The subjugation of the French, and the wresting of all of
their North American territory from them was brought about
through the aid of these colonists. The spirit of dislike, engendered by long years of colonial warfare, was yet exceedingly bitter. In addition, they opposed it on religious grounds.
Hostility to the Roman Catholic religion was everywhere and
frankly acknowledged. The acceptance of the authority of
Rome in the matters of religion, they would refuse, with a determination more definite than they would the interference
of a British King in the realm of politics. From the viewpoint
of these colonists, the Quebec Act gave the Roman Catholic
church a power it ought not to possess, and compelled the
British-born subjects dwelling in Canada, to accept in this
French law and government and religion, a species of rule
which no Britisher would accept of his own free choice.

But another provision of this Act of 1774, displeased the
British colonists in America equally with that which gave
Rome authority in religious matters in Canada. The province
of Canada, when taken over by the British from the French in
1763, included all that territory north of the Ohio, and westward to the Mississippi river, the homeland of many of the
lake-tribe Indians. The Quebec Act set apart this as Indian
territory, the Indian tribes resident there to be considered an
independent nation, and all white men to be excluded from any
ownership or settlement in any part of it. When the westward-
moving colonists stood on the top of the Alleghany mountains,

A PERIOD OF RECREATION

The pioneer mother takes a rest from the multiplicity of her day's toils

and looked northward across this well-watered and magnifi-
cently wooded plain, they saw a country extensive in its boundaries, rich in its resources, with surpassingly fertile soil,
capable of nursing a whole nation within its fertile bosom.
Was this land to be kept a tenantless waste, no white man permitted to develop it, and this after so many of the colonists
had shed their blood to wrest it from France? As the historian
of to-day, reads carefully the chronicles of that period, he is
forced to the conclusion that this provision concerning the
Ohio river territory was one of the chief reasons for the increasing dissatisfaction of the times with Old Country legislation. Here was an enactment that placed itself in direct
opposition to colonial aspiration in respect to the settlement
of the great northwest land. The Indians responsible for the
defeat of Braddock to be given the status of independent
nationhood and the richest territory of the continent to be
their exclusive possession! Their answer was 'No,' and they
made preparation to back their answer with war.

British Colonists Become Enemy Nation to the Motherland

Thus the Quebec Act stirred up many hitherto contented
colonists to find fault with Old Country legislation. Restless
spirits, who had already committed themselves to a policy of
independence, and who had declared themselves in favour of a
complete separation from the Motherland, pointed to the Act,
and declared that, in this measure, there was found ample
justification for their views. It was not long, therefore, until
the British parliamentarians discovered that the Quebec Act,
while it pleased the French, had created a very dissatisfied
spirit among their own British-born subjects in America.

Nine short years after the passing of the Quebec Act, the

British nation entered into a treaty arrangement by the provisions of which she agreed to withdraw from all political
connection with the colonies which she had created in North
America. The French colony of Canada was still retained as
her possession, but shorn of a large slice of its most important
territory. All her lands west of Detroit river, and Lake Huron,
all north of the Ohio and east of the Mississippi extending
northward to the 49th parallel of latitude, these all were handed over with easy grace to the newly-created nation.

Restless Spirit Incited by Quebec Act

But while this settled the matter in dispute arising from
the provisions of the Quebec Act in respect to these colonists,
it did not improve conditions for the British settlers remaining, or those later settling in Canada. The Constitutional Act
of 1791 was passed to remedy the situation created by the Act
of 1774. Let Canada be divided into an Upper and Lower
province, it said; the one province, let it be French in respect
to its people, its government, its religion and its official
language. The other province however, let it be British; its
lands set apart for the British, and its institutions and laws,
British. Separated and independent, let each one work out
its own destiny irrespective and unrelated to the other. Thus
were the waters of the Ottawa made to be a dividing line between two distinct people, as well as the boundary line separating two provinces. It was a policy, in the working out of
which, the features neither of friendship nor of union found
a place, but instead a constantly jarring spirit between the two
races augmented by their differences in language and religion.
Yet not withstanding these racial feelings the statesmen of the
time worked strenuously for unity and harmony. This was

eventually achieved, The French Revolution perhaps having
as much to do as anything in cementing these two distinct
peoples, causing them to seek a united aim in building up the
territory of Canada and making it a country of destiny for both
of them.

An Era of Peace and Good-Will ushered in, 1815

The dominating party in the United States, after union
was effected, made the annexation of all British territory in
North America, their settled policy. This policy was summed
up under the title—Manifest Destiny. Interpreted in the terms
of their desire, it meant that because they wanted the whole of
this country as their own, divine providence had so decreed it.
This would seem to their leaders but a reasonable aspiration.
The men of that generation did not need any other basis on
which to found an expectation for increase of their land holdings other than their ability to carry through their desire for
possession to a successful completion. It was a matter of no
concern to them whether it was professedly the property of
Indians, French, or British Canadians. The only thing which
mattered, was it worth an effort to obtain, and had they the
means by which possession of it could be secured?

When the organization of Upper Canada into a new
province began to be effected, its boundaries were not, at that
time, nor indeed could they be, delimited. In Upper Canada
west, the southern boundary line was the Ohio River; and the
western, the Mississippi. This was the boundary line which
the Forts established by the French and the treaties making
cession of the territory to the British, clearly indicated. When
the Quebec Act, 1774, was being enacted and the territory,

over which the French laws, according to its terms, was to have sway, this treaty designated these two rivers to be the south-western boundary lines of the province of Canada or Quebec. In the meantime, the Treaty of Peace between the United States and Great Britain at the close of the Revolutionary War, permitted a departure from these formerly recognized boundary lines, but subsequent disagreements caused the matter to stand in abeyance for a period of thirteen years. Even then a lack of definiteness about them gave occasion for endless disputations. As a matter of fact, the leaders of the western states could not be satisfied, for their ambition was, that there should be no boundary line, but the whole country the possession of the American Republic. Hence the American invasion of this country in 1812. This war was ended without the necessity of Canadians sacrificing any more of their territory to the United States. Since then, there has been over one hundred years of peace and progress for both countries. An Increasing trade and intercourse has been established and a spirit of good-will has condinued unbroken between them every year since.

As in respect to the boundary line between Ontario and the United States, so also in regard to its eastern and western boundaries separating it from the Provinces of Quebec and Manitoba. These have been finally and permanently settled also. Ontario is now made to comprise an area of 407,000 square miles, with possibilities of development and wealth second to no other province in the Dominion. Essex county comprising its most south-westerly area, contains the home-land of the oldest agricultural settlement in Ontario, and therefore the befitting section where we should begin the study of its achievements in growth, progress and wealth.

CHAPTER III.

The Detroit River Region, Eastward

Its Geography and Topography

I.

The Detroit River District
1701—1815

THE Detroit River District, in early colonial days, was the name given to that section of territory adjacent to and on both sides of the Detroit river. This was the name by which the whole district was known, during the regime of the French and the British. But after the original colony became rent in twain, and the western half of this section of Canada became a possession of the United States, then this name was construed to apply only to that section of the district located in the United States and on the right bank of the river. There still remains, however, the eastern half of the river, the left bank, and all the district adjacent thereto, to which the name also applies. It is with the story of this section of the original district, up to and until the time it became an indisputably British Canadian possession, with which this Volume has to do,—a period beginning with the founding of the French colony here in 1701, and continuing on for more than a century, until the end of the American Invasion in 1815.

Chronologically, Canadian achievement, as it appertains to the province of Ontario, finds its starting point on the banks of the Detroit river. The class of citizens, who comprised the first strata of Canadian nationhood, were the fur-traders, and Detroit was the centre of the fur-trading fraternity. Commencing with the first year of the Eighteenth century, it continued to be the leading centre in the west until after the close of the war of the American Invasion.

The Pioneer Farming Community Established on the Detroit

There were other trading posts established before this one by Cadillac in 1701, but Detroit alone preserved a continuous history and became a permanent part of the subsequent development of the province. These others came, served their purpose as fur-trading posts, and then ceased to be. The Detroit fur-trading post was different. In his conception of it, the founder had in view permanence. So, he included the establishment of a farming community in proximity to it, and as a part of his aim in the founding of it. In both respects his undertakings were successful. The farming community established on the right bank of the river, extended itself in the middle of the century to the left bank also, and became in consequence, the first farming community in Upper Canada. In founding a community of husbandmen, Cadillac showed far-sighted statesmanship. Doubtless, he was impressed with the transitory nature of the fur-trade, and its incapacity to provide alone an adequate foundation on which to build a permanent and prosperous colony. The wisdom of the policy which his prescience dictated, is attested by its subsequent results. It gave to Ontario the first strata of its population, and one which has continued a permanent and important part of the community life of the district ever since.

The Birthplace of a "My Country" Sentiment

But while, in respect to time, the farmer was the first pioneer domiciled in Upper Canada, the fur-trader was the pioneer in respect to the creation of a distinctively Canadian spirit. This arose from the attitude of the Americans to the Indians, and, as a consequence, its effect upon the fur-trade. In those days, when might was right, there was no attempt at effecting achievement by compromise and with a mental attitude conciliatory in spirit. Compulsion was the weapon used both in obtaining the alienation of his lands from the Indian and the settlement of international problems. Had the United States adopted a mild-voiced and a soft-hand policy towards both the Indians and the fur-traders, they would have succeeded in gaining for the Republican idea of government such a spirit of good-will, that Canada, in all probability would have voluntarily followed the example of the other colonies, and have joined them in establishing a confederacy of states for the whole of North America. But, in determining to compel them, by force of arms, to accept the United States programme, they under-estimated the power of the fur-traders, and consequently the power of Canada to defend itself. As a result, when the invasion of the country took place, in 1812, with a view to its incorporation into the Union, the Americans not only failed in their aim, but they left behind a new spirit, or a former embryonic spirit, one stage further advanced in its development. Hitherto, 'My Trade' was the chief interest of these citizens of the West, now they have added another, 'My

Country,' and British connection the distinguishing attribute between it and the United States.

Honored as the First Point of American attack 1812

In the Invasion of 1812, the United States honored the Detroit river district, by making it the first place in the point of their attack. In this district, also, the most telling episodes of the war occurred. The joining of the Indians on the side of the Canadians, was wholly an achievement of the fur-trader. By what occurred in the Detroit river district, the United States gave to Canadian history, three imperishable names. Brock, Tecumseh, Procter. There were many others worthy of a place of equal honour, but these three comprised the leaders responsible for the conduct of the Canadian army during the first fifteen months of the war, the whole period of the Western campaign. Standing on the battle-ground, with their face to the foe, and with their weapons in their hands, the first two of these handed over their lives as a sacrifice for the cause for which they fought, and went out with a halo of glory attached to their name. Procter, after withstanding the enemy in three of the most effective successes of the campaign, met with an equal number of reverses, and went out with the stigma of a great reproach attached to his name. All three fought for an independent Canada, and all three contributed a worthy part to its successful achievement. The death of Brock opened out the door of opportunity for the establishment of another British nation on this continent. The death of Tecumseh closed the door of opportunity for an independent Indian nation. In the destiny of nations, it was decreed that the one should find its place under the sun, adding to the sum

total of world achievements; and that the other should cease
to be, other than a memory.

The Pioneer in respect to Civil Rule

The Detroit river district was pioneer also in respect to
the establishment of Civil rule in Upper Canada. Before this,
all civil cases were tried at either Montreal or Quebec. The
treaty which effected a final settlement between the United
States of America and Great Britain, in 1783, ceded to the
United States all of the Canadian territory west of the Detroit
river, but for reasons which shall later be discussed, this
cession did not become effective until thirteen years
subsequently. During these years, the district was still under
British rule on both sides of the river, though the right bank
was nominally ceded to the United States. It was in this
period that civil rule became an established part of the government of Upper Canada.

In 1788, Lord Dorchester, after having divided the
province into four districts, and giving the name of Hesse to
the Detroit river district, established for it a Court of Common
Pleas, its officials consisting of three justices, and eight
magistrates, a sheriff and a Court Clerk. These officials were
all domiciled in the town of Detroit, and all accepted office
save the three justices. These being laymen, deemed the work
beyond their capacity and petitioned Lord Dorchester that a
trained official be appointed. In accord with this petition,
William Dummer Powell, an able lawyer of Montreal, was
appointed Judge, and thus became the first in line of the
justiciary officials of the province of Ontario. The first sitting
of this court in Upper Canada was held at L'Assumption, since
the town of Sandwich, on July 16th, 1789. Like the part of the

Cadillac colony established on the left bank of the river, it holds, therefore, the rank of pioneer in this field of established Canadian institutions.

The Changing Names of the District

This district, for municipal purposes, was named in 1788, the District of Hesse. In 1792, by proclamation of Governor Simcoe, the name was changed to the Western District. This continued to be its name until the middle of the Eighteenth century, when districts were abolished, and counties and townships became the units of our Municipal system. The boundaries of the district suffered some considerable changes with the progress and development of the country, but finally comprised the area now covered by the three counties of Essex, Kent and Lambton. This volume comprises, therefore, the pioneer history of each one of these three counties.

II.

The Geographical History of
The Western District

AGRICULTURALLY, what was known in the early history of Upper Canada as the Western District, has since been developed into one of the finest on the continent of America. At least it has the possibilities of being the premier, if the potentialities which the resources of climate and soil supply it, are fully utilized. Its total area is estimated as comprising 1,670,500 acres, or 2,610 square miles of arable land.

This district is marked off as a separate area from the rest of the province, by the surrounding waters of Lake Huron, St. Clair river, Lake St. Clair, Detroit river, and Lake Erie, which waters not only form its northern, western and southern boundaries, but give to it as well a peninsular form in shape. It is the most southerly section, not only of the province of Ontario, but also of the whole Dominion. Its latitude is that of France in Europe. It therefore enjoys a climate which supplies conditions for effective farming, second to no other in Canada.

Climate of Western District

"The climate is an exceptionally fine one," reports a government survey of the district. "The winters are comparatively short, moderate, crisp and clear; and the summers, long, warm, and pleasant. Snow seldom falls in sufficient quantity to lie until December, and save an occasional 'November bluster,' in the latter part of the month, fine sunny weather often continues until Christmas. The heat of July and August is greatly modified by gentle winds from the surrounding lakes. Seeding begins from the first to the fifteenth of April, some Springs much earlier; clover cutting commences in the last two weeks of June, and the wheat harvest immediately after. Corn husking is continued through October and November; Fall plowing often long into the winter." In the northern section of the district, the fall of snow in winter and the cold is somewhat greater than in the most southerly. But altogether, the climate is the most temperate in the province.

In addition to the favouring possibilities supplied by the

climatic conditions of the district, the soil on the whole is
unsurpassed for its fertility. Four varieties are found within
the bounds of the district. On the one side there is clay,
heavy, stiff, and impervious, while on the other, as in the
township of Zone, there is a light sand. Between these two,
and comprising about eighty four per cent of the cultivable
soil are loams, varying from a heavy clay to a light sandy
loam. A black vegetable mould is found distributed extensively
more especially on the reclaimed marsh grounds. This
vegetable mould has in some instances a clay, but in others,
a gravelly subsoil. The top soil of the district varies in depth
from four inches to three feet, but generally it is from twelve
to eighteen inches in depth. Little, if any waste lands exists
in any one of the three counties. Practically all of it is capable
of being brought into a state fit for cultivation.

A Level Plain

This area, in topographical feature, is a distinctively level
plain. There are a few elevations and depressions, but these
are so exceptional, or so little marked, that there is in the main
little relief from the monotonous level nature of its superficial
area. As you approach the extreme east or northern boundaries, you come to the outskirts of a rolling or undulating
country, but the peninsula itself, is in the main, free from hill
or dale, save the depressions which basins of water, streams or
rivulets have made.

The Ridge

There is an elevation called "The Ridge," which skirts the
shore in the near neighbourhood of Lake Erie, which is the

one main exception to the level nature of the country. It follows the shore of the lake and in the same direction, now bordering very close to it, and now again some considerable distance from it. In Romney township it is close to the shore line; in the township of Harwich several miles north of it. As in respect to direction, so also in its formation. In some instances, as its name applies, it is a mere ridge. In others, it extends to form a plateau several farm lots in width, with the land south of it, sloping gently to the shore line of Lake Erie.

Depressed Basin between Ridge and Thames

Another feature in the topography of the central section is a depression midway between the Ridge and the Thames river. Here is found a basin of land whose level is lower than the water level of Lake St. Clair, or the banks of the Thames river. The soil of this basin is a heavy clay, and in the early days of its undeveloped state was covered with a heavily-timbered forest mainly of elms, which was brought into a state of cultivation only after a scheme of drainage was inaugurated.

Peculiar Phenomenon of National Drainage

For the development of this peninsula, agriculturally, the greatest problem facing the pioneer settlers, was drainage. There is a very remarkable phenomenon in respect to the natural drainage, or natural slope of the land. The district is traversed by two rivers, both of which flow from the east, westward, and both emptying also into the same body of water, Lake St. Clair. Yet the great natural drainage scheme of this interior country is from the west, eastward. That chain of lakes and rivers starting with Lake Superior, with St. Marys river, Lake Huron, St. Clair river, Lake St. Clair, Detroit river, Lake Erie, Niagara river, Lake Ontario and St. Lawrence river following in order, all form a drainage scheme extending from the west to the east. Yet these two rivers flow in the opposite direction from this main artery indicating a slope from the east, westward, in respect to the district through which they pass.

Thames River

The Thames river, at its source, is made up of two branches which unite at London, Middlesex county, to form one stream, a place designated the Upper Forks in the early period of the country's history. The northern of these two branches, takes its rise in Perth county, and the south in the near neighbourhood of the Grand river. These branches pass through an undulating or hilly country, with gravelly soil, through a territory with a pronounced slope south and westward . . . Their banks are high, their waters swift-running, clear and bright. But after joining together at London, and arriving at the level plain of the Western district, the river becomes almost devoid of banks, its current sluggish, and its waters murky and gray, these features becoming more pronounced, the nearer it approaches its outlet.

Sydenham River

The Sydenham river is also made up of two branches, which join together at Wallaceburg. It takes its rise from the same watershed as the Thames, but further north, and flows westward parallel with it, and empties into the same body of water, Lake St. Clair. The same general description applies to it as to the Thames river, the same lowness of banks, the same sluggishness of current, and like it, growing more pronounced the nearer it reaches the great interior water system into which they both empty themselves.

The Want of Slope for Drainage

How gentle is the slope, and how level the surface of the plain of which these two rivers are the main drainage system, may be gauged from considering the water levels of the two lakes, Huron and Erie. Lake Erie is five hundred and seventy three feet above the sea-level, but Lake Huron only five hundred and eighty one and a half. This leaves a slope to the land of this peninsula from north to south of less than nine feet, or a general fall of about one and one-quarter inches per mile. As the early explorers viewed this land, they adjudged that only the portion bordering on the banks of these two streams was fitted for farming purposes. The rest, they concluded, was too wet and low-lying, and devoid of sufficient slope, to make it capable of cultivation. This would be quite true, if the drainage of the country was confined to the system which Nature alone provided. But with the removal of the forests and the achievement of an artificial drainage system which covers the whole peninsula, there is to-day not an acre of the whole plain but which provides an opportunity for successful crop-growing.

Navigability of Both Rivers

Both of these rivers are navigable rivers in a measure. The Sydenham, however, promises to be the greater asset of the two in respect to transportation. The Thames river has a sand-bar at its mouth, a constant menace to safe navigation. In addition, though navigable to Chatham, its depth of water is too shallow to permit of anything other than vessels of light draught to pass up the river, and there is not sufficient commerce to justify the expenditure necessary to keep the channel clear from the constantly recurring menace which the sand-bar occasions. The Sydenham, has, however, a depth of fourteen feet of water from its outlet as far east as Wallaceburg, and is navigable for light craft as far as Dresden. As a result of this more advantageous circumstance, ocean freight vessels are now making Wallaceburg a port of call in connection with the sugar-refining industry established there, carrying raw sugar from the West Indies to Wallaceburg, as the inward freight, and for return cargo, the finished product, the refined sugar, to be freighted to its appointed market.

There are, in addition, to these two rivers, numerous other streams and creeks, of which the Aux Sable at the north, and McGregor's creek in Kent county are among the largest. The Aux Sable takes its rise from the same watershed as the Sydenham, and flowing northwestward, empties into Lake Huron. McGregor's creek takes its rise at the eastern boundary of the district, drains the basin of lowland between the Ridge and the Thames river in the townships of Howard and Harwich, and then turning northward, empties into the Thames river at Chatham. Numerous other rivulets, some empty themselves into the main artery of drainage, while others of them are tributaries of the Thames and Sydenham river.

Artificial Drainage an additional Necessity

Ordinarly, these streams would be considered sufficient to effect an adequate drainage of the district, yet, owing to the

OLD WINDMILL

This historic windmill was a landmark for many years
of pioneer conditions on the Canadian Detroit river
frontier. Owing to the level nature of the district and
the consequent absence of streams for water-power the
thrifty early French settlers overcame the difficulty by
erecting wind-mills to supply the motive power to grind
their wheat and corn into flour. These were erected at
convenient distances apart along the river's bank by
the first settlers. The above, the Montreuil mill, was
situated on the shore above Walker's distillery and as
the last to be preserved of these ancient industrial plants,
it became the subject of both comment and portrayal by
many artists and writers, and has in this way, been a
medium to give increased prominence to this worthy and
well-known of the early French Canadian families, representatives of whom still live on the pioneer homestead
lands.

level nature of the plain, as has been observed, and the lowness of the land bordering on the main artery of drainage, notwithstanding the number of them, they have been found to be quite inadequate to supply a sufficient drainage to insure effective farming operations. Hence, an artificial drainage system had to be inaugurated. This system divided itself into two parts. Where the land was high enough above the level of the waters of a natural stream or outlet, nothing was required other than to provide a drain of sufficient dimensions to carry off the surface water of the section of land through which it passed. But in respect to those tracts of land lower than the water levels of the natural stream, a pumping system had to be installed. In this way, 30,000 acres of marsh lands in Raleigh township and areas bordering on the shores of Lake Erie have been reclaimed, and made into rich and fertile farms. Thus by linking up the artificial with the natural, a drainage system has been established, which has transformed the district from a low-lying and undesirable area to one of the most excellent agricultural communities on the continent. In respect to agricultural undertakings, there has been no achievement or activity which surpasses this in importance, and in its production of adequate and desirable results.

Crop-Growing

The development of the capacity of the varied soils of the district for the production of abundant harvests, and the determination of the class of farm products which each variety of soil is best adapted to grow, are as yet in the experimental stage. All the varieties of grain grown in other sections of the province can be grown here also. In addition, there are crops which are staples in this district which cannot be grown in other sections less favoured with climatic conditions.   Tobacco, corn and beans among grains; tomatoes, melons and onions among vegetables; and peaches, grapes and apricots among fruits; these comprise nine varieties of farm products which the soils and the climate of this district make especially productive and profitable.   The clay loam is well suited for the growing of sugar-beets and corn, as well as wheat, barley, oats and all other varieties of coarse grain. The soil of the Ridge, being a light loam, is well suited for the growing of certain varieties of tobacco.   The sandy loam in the near vicinity of the Thames river, overlaid or mixed with a rich vegetable mould is exceedingly fertile, easily worked, making farming on these lands both an easy and profitable occupation.   There are some sections in which sand in the loam too greatly predominates, which makes farming only to those who employ scientific methods and the generous use of artificial fertilization, a profitable and successful occupation. Some of the clay lands require excessive labor to cultivate, but are well-suited to the growing of hay and clover, and consequently are better devoted to pasturage and stock-raising.   The season for work on the land is much longer in this district than other sections of the province, which is an added feature of favourable circumstance for an increase in the productivity of the soil, an additional aid to place the district in the front rank among the agricultural communities of the province.

Annual Over-flow of River Banks

One other feature, arising from the low-lying nature of the land, and the low banks of the rivers, remains to be mentioned. This is the inundation of a large area of the district with the melting of the snow every approaching spring. This is especially true of years when the snow fall has been heaviest and has remained unthawed until the early Spring. There does not seem to be any remedy for this, as the conditions which produce it are climatic and topographical. The annual snow-fall is heaviest in the London region, and the territory drained by the north branch of the Thames. The numerous artificial channels, some of them but ordinary drains, others of them of dimensions sufficient to place them in the class of small rivers or canals, carry the surface water very rapidly to the main outlets. There are no high banks westward to hold any great quantities of water. Hence the inundation of the surrounding region is inevitable, the greatness of the overflow being determined by the amount of water carried down. The danger is aggravated, when the ice has not left the river, ice-jams being oftentimes formed, which make an effectual water-stop. And the inundation of the adjacent land an increased certainty. The seepage carried down the river during those Spring freshets, and spread over the land, is said to increase its fertility, and thus provides a compensating good for the inconveniences arising from this annual overflow. The damage to property is not so great as it would be, if the current of these rivers was more rapid. Still it remains, however, an unfavourable and inconvenient annual handicap, which all would gladly remove if there was any way commercially possible to do it.

The removal of the forests and the drainage of the district is an achievement which has been accomplished within the memory of men now living. Another century of equal progress, should without fail lift the district to the front rank of rural districts any where in the continent. Nature has supplied the soil and the climate. It only remains for the district to supply the human enterprise required to successfully achieve it.

Mineral Productions of Ontario

In accord with its development, Ontario is divided into two sections, one called Old and the other New. What is termed New Ontario has been slower of development because it is fitted only for mining and lumbering industries The metals found are gold, silver, copper and nickel and are widely distributed. Mining in many of these places has been found commercially profitable. The pulp industry of the north has also attracted many settlers. These two industries promise to revolutionize our former conceptions of this north country. What is termed Old Ontario is divided into two sections, by an elevated range of limestone rock beginning at Queenston Heights on the Niagara River and extending north-westward throughout the Bruce Penninsula. The tract southwest of it comprises nearly 15000 square miles. At the very south-westerly section is a level plain, but north-easterly, rolling hills of boulder clay or stony moraines come into increased prominence. In the lower levels throughout this tract, plains are found gently sloping towards the nearest of the Great Lakes but as you go northeastward, the rolling, hilly feature of the tract becomes more pronounced. In contrast to New Ontario, no precious metals are found in this section of the province. In the Detroit region petroleum, salt and natural gas are discovered in considerable quantities and provide an important industry for the district.

CHAPTER IV.

The Early French Settlement

I.

WOOD-RUNNERS
(Coureurs de Bois)

Their Contribution to Canadian Achievement in the Field of Exploration.

THE statement attributed to the historian, Bancroft, that "Scarce a cape was turned, scarce a river entered, but a Jesuit led the way," can hardly be accepted as historically correct. The first of the Indian missionaries to arrive in Ontario, Joseph Le Caron, was not a member of the Jesuet order, but a Recollet priest. The Jesuits have had a wonderful history to their credit. Their pioneer activities have been marked by a remarkable display of intrepid courage and self-sacrificing zeal. It would indeed be difficult to exaggerate the place they held in the pioneer life of the country. Yet, notwithstanding the importance of that place, they were not the first discoverers and explorers of the land. That place of honour belongs to that class of Frenchmen known as coureurs de bois, or wood-runners.

Men of Adventure

These were adventurous men, full of energy, daring and enterprise, who came into the country as traders, trappers and hunters. They entered into the wild life of the New World with zest, attracted to it by the opportunity for adventure and freedom which it provided, as well as the profit in furs which it supplied, a trade which assumed, in the course of time, an enterprise of great magnitude.

Thousands of these men went into the woods during the first decades of our country's history. When the profits in furs came to the ears of the people in France, young men, representing all grades of social life, inspired by the stories which they heard, set out to make their fortunes in Canada. They did not wait for the Indians to bring their furs into the market, but struck out boldly into the hunting grounds for themselves, some to dicker with the Indians, giving in exchange trinkets of no value whatever, others to hunt and trap on their own account. Powder and shot, brandy and rum, were the chief commodities of trade. The first two were required for hunting purposes, and the last two to satisfy a thirst for intoxicants on the part of the Indians which was carefully cultivated by the traders.

Their Descendants

The coureurs de bois learned the Indian language, adopted the Indian mode of life, and married Indian women. They did not take part in any way in the war raids of the Indians, but gave their undivided attention to their own chosen calling. Some of them became absorbed in the Indian tribes, but others maintained their own individuality, their progeny growing up into a distinct race to whom was given the name, Canadians, or Half-breeds.

The First Discoverers and Explorers of the Country

These men were the real discoverers of the country, and it was from the knowledge obtained from them, that the men whose names have been handed down in history as pioneer explorers, received their inspiration to seek, and guidance to discover, the extent and the possibilities of the country. A representative of the coureurs de bois was guide to them on every new discovery. Also, it was from these that the first Jesuit missionaries derived their knowledge of the Indians and the location of the Indian villages.

They have left no records of their discoveries, the dangers which they faced, the journeys which they achieved, and the sacrifices which they made. They did not belong to a class who write, or who take the leisure to write, concerning the things which they had observed or achieved. Happy-go-lucky, they were at all times ready for a new adventure. Their life was one of continuous romance, yet they never recorded a detailed story of their active and dangerous experiences. They went into the woods with their guns, axes, knives, and trinkets. Some returned; others did not, but became victims of accident, disease, or the treachery of hostile Indians. Those that came out of the woods brought their packs of beaver skins and other peltries worth hundreds of crowns, and sold them to the merchants in Montreal or at the nearest trading post.

A Tax on Their Trade

At first the coureurs de bois were permitted to undertake their enterprises and carry them out untroubled by any overruling authority. But this freedom was of short duration. It was not long until the French crown demanded a share in the profits, and only those were permitted to trade in furs who had purchased a license from the ruling sovereign or his representative. In the first instance the merchants were the purchasers of these licenses. They received them from government-appointed officials, and in turn employed the coureurs de bois as their agents. The officials of the government looked to the merchants to share with them the profits of the trade, so that the crown, the officials, the merchants, and the coureurs de bois made up a line four-deep with hands stretched out to the Indian, begging for furs and profits.

The Indian Exploited for Gain

It can easily be seen that where so many profits were to be taken, either the Indian would get very little for his furs, or the ultimate purchaser would have to pay an exorbitant price. The inevitable result followed. The coureurs de bois, compelled to provide profits for themselves and their superiors, showed no conscience whatever in their dealings with the Indians, but cheated them on every occasion. As was to be expected, also, ways and means were discovered for evading the license tax, the price of which was six hundred crowns. This permitted the lading of only two canoes with supplies for barter. By permitting the coureur de bois to exceed the conditions of his permit, allowing trade and traffic amongst unlicensed persons, and by other clandestine ways, the government officials and merchants assisted in robbing the crown of its expected dues, and at the same time added to their own profits. Whether officials, merchants, or coureurs de bois, none of them seemed to be above taking advantage of a situation which would provide increased gain to themselves without regard to the claims of the crown.

Their Morality Equal to the General Standard

There is nothing to indicate, though his character has been frequently painted in sombre colors, that the morality of the coureur de bois was below the general standard of the rest of the colonists. The lack of sobriety and chastity, characteristic of his class, was but an expression of the conditions of the times. He loved the canoe, the swift running rapids and the deep woods, and by means of these he obtained his adventure and made his living. To them belongs the honor of discovering this district, and two of them, we know, had a trading centre established on the Detroit River before any other attempt was made to establish a settlement there.

The story of these wood-runners should occupy the second chapter in a consecutive history of our country. How interesting that chapter would be, could it only be written. But no record of their activities having been left us, we will have to content ourselves with such gleanings of the story as the reading of the subsequent history of the country will supply us. To the constructive imagination belongs the task of putting these together, and supplying what otherwise would remain as an irremediable loss to the literature of our country.

II.

THE CADILLAC COLONY

An undertaking out of which there has resulted the erection of a great city on the banks of the Detroit river, and the establishment of the first Agricultural Community of Ontario.

THE first attempt at founding a permanent settlement on the Detroit river was undertaken by Cadillac, a French army officer of energy, prescience, and ambition. He was a native of Gascony, France, whose marriage to Marie Therese Guyon took place in 1687 in Quebec, he being then twenty-six years of age. After several appointments under the government in the Marine Department and at other places, he became commandant of the French military post at Michilimackinac in 1679. While there he conceived the plan to establish a military and trading post on the Detroit. This was ostensibly to prevent the encroachments of the British on the French territory and the French fur trade, though some claim that his aim was only to further his own personal ambition.

Opposition to Cadillac's Enterprise

His project was opposed from many quarters. It was opposed by the Jesuit missionaries on the ground that it would tend to draw away the Indians from the Michilimackinac post, and so injure their cause and weaken their hold upon the Indians there. It was opposed by the governor on the ground of expense. It was opposed by the merchants of Montreal as tending to draw away traffic from the well-established trade route by way of the French river, Lake Nipissing and the Ottawa river to Montreal, and substitute for it the shorter route to Albany and the British market. The Albany merchants, the Indian was already coming to know, supplied him with cheaper goods to buy and paid him a higher price for his furs, than the Montreal people did. Notwithstanding all the forces opposed to his projected undertaking, Cadillac succeeded in obtaining the good-will of his sovereign. Permission was given him to establish a post there with the promise of a grant of land fifteen arpents square. "Wherever on the Detroit river the new fort should be located."

"His most inveterate enemies were the Jesuits and the Company of the Colony of Canada. The quarrel with the Jesuits was of long continuance, and was part of the troubles that started between that order and Governor Frontenac. Cadillac would not permit the Jesuits to establish themselves at Detroit, and the church here was under the supervision of the Recollet order.

"The Company of the Colony was not pleased that Cadillac had the exclusive right of trading at this post, and they set about to ruin Cadillac and destroy his village. All parties appealed to the law courts and to the king, but Cadillac was temporarily victorious. In the end they succeeded in driving him from Detroit, and his successor took all his property, estimated at 50,000 livres, and refused to account for ot.*

A Farming Community Part of the Plan

Cadillac's plan not only included a military post with a strong garrison, which would protect the entrance into the lake region against the encroachments of the British traders, but in addition, he contemplated planting an extensive colony of permanent settlers, to clear the land and till the soil, who in the course of time would be able to supply the post with all of its needed provisions and thus make it self-sustaining. Appointed by the king as commandant, Cadillac arrived in the midsummer of 1701, and on the 14th day of July made selection of the site, where, with fifty soldiers and an equal number of artisans and traders, he began another chapter in the history of French Canada. The new post suffered varied vicissitudes of fortune in the early years of its history, but in the course of time all difficulties which hindered its progress were overcome, and to-day the magnificent city of Detroit is the final realization of Cadillac's vision, idea and ideal.

---

*Early Detroit by C. M. Burton, 1909.

Cadillac* was born in Gascony, France, about 1656. "He came to America in 1683, and settled at Port Royal, now called Annapolis, near the home of the heroine of Longfellow's "Evangeline." He became a man of considerable importance to the French Government in consequence of his knowledge of the New England Sea Coast . . . He had drifted to Quebec within the next few years, and there he married Marie Therese Guyon, June 25th, 1687. In 1688, he received as a grant from the French Government, Mount Desert Island, and a great tract of the mainland opposite the island, including the whole of Bar Harbour, Maine.

"He was at Mackinac as Commandant from 1694 to 1698, and then, passing through Quebec, he went to Paris in order to lay before the king his project for establishing a colony at Detroit. Successful in his errand he returned to America, and started from Montreal for his new home on the Detroit river, on June 2nd, 1701. He was accompanied by one hundred Frenchmen and one hundred Algonquin Indians.

Cadillac remained the commandant of Detroit for ten years, and until the year 1711. He did not always get along well with his own people. He was opinionated and quarrelsome. Those who lived in the village were compelled to do as he directed, but some of them had influential friends in Montreal and Quebec, who took up their quarrels, and finally succeeded in having Cadillac removed.

"Later in life he was appointed governor of Castell-Sarrazin, in the southern part of France, and died there October 18, 1730. He never recovered his losses at Detroit, but the State of Massachusetts gave to his grand daughter, Madame Gregoire, his old land grant at Bar Harbor . . . "

---

*Early Detroit, by C. M. Burton, 1909.

CHAPTER V.

The Coming of the British

THE long drawn out struggle between Great Britain and France came to an end in 1759 by the final defeat of the French army at Quebec, and the surrender of the Fort and country to the successful British. In all of this struggle the colony at Detroit suffered nothing of the hazards of war. It was far removed from the recognized boundary line separating New France from New England. But the isolation which hitherto had supplied them with comparative safety was now destined to be disturbed. In the early days of November, 1760, Indian scouts brought them the news that a flotilla of fifteen boats and two hundred British soldiers were making their way along the shores of Lake Erie, and heading apparently for Detroit.

Major Rogers despatched to Detroit

It was Major Rogers, sent from Quebec to receive formally the capitulation of Detroit and the other French outposts in this western country, and with him a small contingent of rangers to take up garrison duty in these surrendered forts. In a post so important as Detroit, and with traders passing to and fro constantly between it and Montreal, news must have reached them of the British conquest and the surrender of New France to the victors. Doubtless, also, the garrison would be in expectancy of a British expedition being sent to their

country to take possession and occupy the centres of trade
hitherto owned and occupied by the French.

Difficulties of the Journey

The expedition took much longer to make the journey
than they had anticipated. They had set out from Montreal
before the middle of September, choosing the southern
route by way of Lakes Erie and Ontario. Unfortunately for
their comfort and the progress of their journey, they en-
countered a series of severe storms making the navigation of
their flotilla both difficult and dangerous. The month of
November had already set in before they reached the western
limit of Lake Erie. Arriving there, they found themselves
in Indian country, with Colonel Belestre, the commandant
then at Detroit, apparently not intending to surrender the
post unless there were official orders from Governor Vaudreuil
in possession of Major Rogers, the commander of the expedition. Under Belestre's instructions, it is presumed, Major
Rogers was met by several contingents of Indian warriors, who
were prepared to prevent his further approach towards the
fort, if the occasion required it.

The Ottawas stand in His Path

The first of these to arrive was a contingent of the
Ottawas. This was on the seventh day of the month. They
were followed a short time after by their chief, Pontiac. It
was late in the day, and Major Rogers was in the act of preparing an encampment for the night at the mouth of the
Grand River, when he put in his appearance. "It was here for
the first time," wrote Parkman, "that this remarkable man

stands forth on the page of history. He greeted Rogers with the haughty demand, what was his business in that country, and how he dared enter it without his permission."

Major Rogers adopted a conciliatory attitude, informing him that the country had been surrendered to the British, by the French, and that they had come to obtain a peaceable possession of it. He solicited the good-will of the Ottawas and agreed to establish a friendly alliance with them.

How was Pontiac Conciliated?

In making preparations to come to this distinctly Indian country, had Major Rogers brought with him gifts, expressions of friendship to conciliate the Indians and win them over as allies of the British, in the same way as they were before, the friends and allies of the French. Was it sufficient for him to say, "The French are defeated; the British are victorious?" We do not think that statements similar to these, no matter how strongly made, would have enough of influence over the Indian to change him, from a possible enemy, to a faithful friend. Hitherto, it was gifts that purchased the friendship of the Indian, not words of blandishment, nor of bravado. In Indian Councils, the successful treaty maker would say, "Your great Father across the sea is strong; he has defeated the French; but he loves the Indian, and he has sent to you a token of his love—a piece of tobacco, a glittering trinket, a gill of rum." Had Major Roberts brought with him those requisites which would enable him to talk to the Indian in terms of gifts? Pontiac came; considered the proposals of Major Rogers, and went away conciliated, if not permanently, for the time being at least. Four chiefs of the Wyandots and with them four hundred of their followers, and they too went away conciliated. Four hundred pieces of tobacco, four hundred glittering pieces of glass, and four hundred gills of rum would have far more weight in obtaining a safe passage up the Detroit river than would the drawn swords of two hundred British soldiers. Having in mind the predominant trait of Indian character, his weakness for gifts, we think it would not be a matter of great surprise to be informed that as a part of the cargo of the fifteen boats which comprised the vessels of his expedition, there was supplied by Major Rogers those things which would enable him to add friendly works to friendly words in his negotiations with the Indians.

This appears all the more likely, since it is known that Pontiac had previously visited the British at Pittsburg enquiring "what treatment they would have, should the English succeed, to which was answered that first all the rivers were to run in Rum, that presents from the Great King were unlimited, that all sorts of goods were to be in the utmost plenty"*

An Embassy from the French Commandant

Having conciliated the Indians, whatsoever the method by which it was effected, Major Rogers continued his journey up the Detroit, and was honored on the way by a third embassy. This was from the Commandant of the Detroit post, requesting him to forward a copy of the capitulation papers, and the letter of instruction, which he had from the French Governor, respecting the capitulation. Major Rogers, anticipating this demand, had some time previously despatched Lieutenant Brehme, to inform Colonel Belestre of his arrival, and his possession of these necessary official papers.

---

*Michigan under British Rule, Riddle, P. 400.

BOIS BLANC ISLAND

Situated at the mouth of the Detroit river, opposite the town of Amherstburgh. In its early history it was claimed and occupied by a tribe of Wyandot Indians; Father Richardie established his first Canadian Mission there; it was the chief place of encampment for the Indians, assisting the Amherstburg garrison in its defence against the American invaders, 1812-13. It is now, 1929, established into a popular pleasure resort, by the Detroit and Windsor Ferry Company.

In order to further assure himself of the certitude of these statements, he wisely forwarded this embassy to receive a copy of them. In accord with this request, copies were forwarded under the care of Captain Campbell, a trusted officer, whose subsequent experiences at the Fort supplies a tale of tragic and fateful import in connection with the Pontiac conspiracy.

The British Camp on the Eastern Shore

On the 28th, the expedition arrived at its destination, but Major Rogers delayed the formal acceptance of the capitulation, until the following day. When he came within near reach of the town, he turned to the right and prepared his encampment on the eastern, now the Canadian side of the river. It seems prophetic that he should have done this, as if presaging the future history of the country, that on this shore, and not on the west, the British domain would be permanently established.

The Fleur-de-lis Lowered from the Flagstaff

On the morning of the 29th, the capitulation was duly received by a detachment of Major Rogers' men, under the leadership of Lieutenants Leslie and McCormick. The rest of the expedition was half a mile from the fort when the transfer took place. In the presence of the French who comprised the population of the fort and representatives of the colonists from both sides of the river, and of the four Indian tribes—the Hurons, the Ottawas, the Pottawatamies, and the Ojibways— the Fleur-de-lis was lowered from the flagstaff, and the British flag made to take its place.

The rule of France on the Detroit river had now come to
an end, but not before the colony, which had been planted
there, had been established on a permanent basis. The French
soldier departed, but not the French habitant and the French
trader. These remained and took a prominent place in the
subsequent history of the district.

Francis Parkman Describes the Historic Incident

"Rogers had now entered the mouth of the River Detroit,
whence he sent forward Captain Campbell with a copy of the
capitulation, and a letter from the Marquis de Vaudreuil,
directing that the place should be given up, in accordance with
the terms agreed upon between him and General Amherst.
Beltre was forced to yield, and with a very ill grace declared
himself and his garrison at the disposal of the English
Commander.

The whale boats of the rangers moved slowly upwards
between the low banks of the Detroit, until at length the green
uniformity of marsh and forest was relieved by the Canadian
houses, which began to appear on either bank, the outskirts
of the secluded and isolated settlement. Before them, on the
right side, they could see the village of the Wyandots, and on
the left the clustered lodges of the Pottawatamies; while, a
little beyond, the flag of France was flying for the last time
above the bark roofs and weather-beaten palisades of the little
fortified town.

The rangers landed on the opposite bank, and pitched
their tents upon a meadow, while two officers, with a small
detachment, went across the river to take possession of the
place. In obedience to their summons, the French garrison
defiled upon the plain, and laid down their arms. The fleur-de-

lis was lowered from the flagstaff, and the cross of St. George rose aloft in its place, while seven hundred Indian warriors, lately the active allies of France, greeted the sight with a burst of triumphant yells. The Canadian militia were next called together and disarmed. The Indians looked on with amazement at their obsequious behaviour, quite at a loss to understand why so many men should humble themselves before so few. Nothing is more effective in gaining the respect, or even attachment, of Indians than a display of power. The savage spectators conceived the loftiest idea of English prowess, and were astonished at the forbearance of the conquerors in not killing their vanquished enemies on the spot.

It was on the twenty-ninth of November, 1760, that Detroit fell into the hands of the English. The garrison were sent as prisoners down the lake, but the Canadian inhabitants were allowed to retain their farms and houses, on condition of swearing allegiance to the British crown. An officer was sent southward to take possession of the Forts Miami and Outanon, which guarded the communication between Lake Erie and the Ohio; while Rogers himself, with a small party, proceeded northward to relieve the French garrison of Michilimackinac. The storms and gathering ice of Lake Huron forced him back without accomplishing his object; and Michilimackinac, with the three remoter posts of St. Marie, Green Bay, and St. Joseph, remained for a time in the hands of the French. During the next season, however, a detachment of the 60th regiment, then called the Royal Americans, took possession of them; and nothing now remained within the power of the French, except the few posts and settlements on the Mississippi and the Wabash, not included in the capitulation of Montreal.

The work of conquest was finished. The fertile wilderness beyond the Alleghanies, over which France had claimed sovreignity,—that boundless forest, with its tracery of interlacing streams, which, like veins and arteries, gave it life and nourishment,—had passed into the hands of her rival. It was by a few insignificant forts, separated by oceans of fresh water and uncounted leagues of forest, that the two great European powers, France first, and now England, endeavoured to enforce their claims to this vast domain. There is something ludicrous in the disparity between the importance of the possession and the slenderness of the force employed to maintain it. A region embracing so many thousand miles of surface was consigned to the keeping of some five or six hundred men. Yet the force, small as it was, appeared adequate to its object, for there seemed no enemy to contend with. The hands of the French were tied by the capitulation and little apprehension felt from the red inhabitants of the woods. The lapse of two years sufficed to show how complete and fatal was the mistake."

The Change of Authority, a Change of Policy

The transfer of the flags meant more than a change in authority. It was an emblem also of a change of policy in respect to the development of the Detroit and surrounding district. The British settler came into the country seeking farms, not furs. The policy of the French meant the increase of trade at the expense of the resources of the country. Every hundred beaver skins meant a hundred beavers less in the woods. Let this be carried on more rapidly than production, and soon beavers would cease to exist in the land.

The policy of the British was increase of wealth by increase of resources. One hundred acres of land growing trees and beech-nuts—let that hundred acres grow corn and wheat, cattle and children instead. Soon there would be ten white warriors for one Indian. The policy of the one added settlement to settlement, making one completed and united whole, increasing in strength as it increased in size. The policy of the other created segregated units, placed them far distant from one another, a policy well suited to the fur-trade in time of peace, but increasing their hazards in time of war. The policy of the one conquered; the policy of the other brought defeat and subjugation. The laws for nations are as the laws for individual men. He who would end triumphantly must husband his resources and increase them.

The French policy did neither. The forests, lakes and streams provided their annual harvests without any labor of man required to produce them. This storehouse of Nature, they returned to it, season after season, deeming that as in the past, so in the future, it would continue to supply its undiminished quota of beaverskins. When the wild life of one locality became exhausted by reason of the unremitting slaughter prosecuted for the sake of their pelts, the hunter trecked further afield until he obtained another more prolific. Increase of trade meant increase of slaughter. No great or permanent colony could be built up on the fur-trade alone, and this was the only vocation which the French seriously followed. Hence their increase in numbers was not commensurate with that of other nations which took seriously to husbandry, a circumstance which led eventually to their defeat in war by the British.

In the same way, with the increase of the number of their forts, made necessary by the lengthening of their frontiers with every acquisition of new territory, they lessened the man-power defending each one, and so weakened the whole front line of their defences.

The French Hope of a Re-Conquest*

But if the British deemed that to them had come a permanent victory, not so the Frenchmen of the Detroit. They knew that a catastrophe had happened to the French arms, but they considered this only in the form of a temporary truce, one of those unavoidable incidents in all warfare, but by no means bringing about the end of their colonization work in North America. In a short while Old France would recover herself, strengthen her position, and with increased vigour take up the cause of her colonists and place them once more on the pedestal of success and power. This they believed, and this also they preached to their old-time allies, the Indians. The Treaty of Paris had not yet been signed, and even when it was signed, it was months afterwards before the habitants of Detroit had been made acquainted with its terms. In the meantime the country was being defended for the British by a few handful of soldiers and these scattered far apart. Would it not be easy to attack and destroy them all, and thus pre-pare the way for the expected return of French power? In the three years which was to intervene before the Treaty of Paris had disabused their minds of these vain hopes, they succeeded in maintaining a first place in respect to their friendship and alliance with the Indians, while at the same time there was fostered among the Indians a spirit of hostility towards the British garrisons now stationed in the West.

*See Anonymous Diary of Pontiac Conspiracy published by C. M. Burton, Detroit.

CHAPTER VI.

Indian Tribes of the Detroit Region

URING the period of the French regime in Canada, that portion of the country west of the Ottawa river was left an unpeopled wilderness, save the one colony established in 1701 at Detroit. The French nation prized this country because of the prolific harvest of furs which it annually produced. Respecting its agricultural possibilities, they had no concern. They established fur-trading posts at strategic points but only among Indians who were friendly disposed towards them.

Michilimackinac, First French Post Established in the West

In the early history of their fur-trade, they established a post at Michilimackinac and made this the chief centre of their western trade. This was because the way of travel from Montreal to the Indian country of the great lakes, had to be the northern route, up the Ottawa river, then west by way of the Mattawa river, Lake Nipissing, the French river and numerous portages between these to the Georgian Bay. They had no alternative other than to choose this route because of the hostility of the Five Nation Indians who peopled the Lake Ontario and the Lake Erie region. No flotilla of canoes could expect to escape attack from representatives of the Five Nations if they attempted the southern route. But in course of time, a treaty of friendship was established between the French and the Five Nation Indians, and then the southern route became opened for trade with Montreal.

Cadillac seized this favorable circumstance to establish a fur-trading post on the Detroit river, the very centre of the greatest fur-bearing region on the continent. With its establishment, the importance of the post at Michilimackinac waned, while that of Detroit proportionately increased. Accompanying him to the place were one hundred Indians, who were the first of many contingents making their way from the north, to transfer their place of abode to the Detroit region.

The Neutrals

The Indians which tradition first attaches to the Detroit region were the Neutrals. These, however, suffered a fate similar to the Hurons at the hands of the same tribes—the Iroquois, or Five-Nation Indians. The extermination of the Neutrals was as complete as was that of the Hurons in the Lake Simcoe region. The few who escaped, became absorbed in the other tribes and thus lost their tribal identity. A tribe of the Senecas took possession of the vanquished territory of the Neutrals, and established a village on Lake St. Clair, near the mouth of the Thames river. But they were not allowed to hold for any lengthened period the lands they wrested from them. They were attacked by a combined army of Ottawas and Chippewas, who meted out to the Senecas, the same treatment which they had measured out to the Neutrals. Thus by migrations from the north, and by conquests over the Iroquois, the Ottawas and Chippewas, the Hurons and the Pottawatamies became the occupants of the Detroit river region during the

French regime and for some considerable time after the arrival of the British there in 1760.

The Four Tribes, Algonquins

These four were kindred tribes and belonged to the great Algonquin family, a race which practically comprised the whole of the Canadian Indians. In the early colonization period of Canadian history, the Indians were divided into two family groups—the Iroquois and the Algonquins. The Iroquois occupied the country south of Lake Ontario, wholly situated in what has since become United States territory. The Algonquins were scattered over a wide area, extending all the way from the Atlantic to the west of Lake Superior, and were divided up into many tribes of which these four were the principal ones.

There are some who claim that the Hurons and the Iroquois belonged to the same family. Others, with more of a show of reason, claim that in the similarity of their language in its root formations, and in their ability to understand each other, each speaking its own dialect, there is ample evidence that the four were kindred. Whether or not of the same blood relationship, tradition claims that it knows of no time when these did not stand on terms of friendship to one another. "No club was ever raised by either of these against the other," is the tradition which goes back as far as the Indian memory.

Their former Homelands

When the French took possession, the Hurons were occupying the country south and west of the Georgian Bay. They had well-populated and numerous villages in that section especially lying between Lake Simcoe and the Bay. The

Ottawas were more nomadic and scattered. Some dwelt on the Manitoulin and other Lake Huron islands; some on the North Shore, and some also at Green Bay, Michigan. The Ojibways or Chippewas, were even more scattered. Villages of them would be found anywhere along the North Shore of Lake Huron, on both sides of the St. Mary's river, and on either side of Lake Superior, while a considerable number of them were located on the north-west side of Lake Michigan. The Pottawatamies were located on both the west and east side of Lake Michigan, wholly in Michigan territory. When Cadillac founded his post on the Detroit, a goodly representation from each of these tribes came and settled in the near neighbourhood.

Birch-bark Canoe Exhibits Inventive Skill

In general characteristics, these four tribes bore a marked resemblance to each other, their differences, if any, being only a matter of degree. None of them were without evidences of ingenuity and skill. Their chief invention was probably their canoe. This was made of birch bark, which was carefully taken off the trees in large squares, and sewn together with thongs, the seams being carefully pitched over with the gum obtained from the pine and other evergreen trees. The ribs were made of cedar, this also like the bark, to insure lightness, a quality of no small consideration where it was so oftentimes necessary to carry them when portaging from one lake to another, or to avoid a rapids or fall in the river. In every feature it was well adapted for the purpose, an evidence of their inventive genius and their skill in manufacture.

Many of their household utensils, also, were made out fo birch-bark, but some, however, out of wood. Their sap-troughs and buckets were all made of birch-bark and ,constructed on the same principle as their canoes. Vessels of pottery, chiefly bowls, were found in use amongst them at the time of the advent of the whiteman. They also made use of copper for the manufacture of tools and utensils, which they hammered into shape, knowing nothing of the principle of melting metal.

Their System of Tanning Skins

Their second greatest discovery or achievement, was their method of tanning the skins of wild animals, from which they made their moccasins and their clothes. In a sense this may have been their greatest since it was so great and universal a necessity. In the manufacture of their moccasins and clothes the women evinced much taste as well as skill, and the Ottawa women are said in this to have excelled all others. Their ornaments, made with porcupine quills dyed in various colors, displayed the qualities both of skilful art as well as beauty in design. Their baskets made from woven ash fibres, their household vessels made from birch bark, their mats made from woven grasses, all gave evidence that they lacked neither in industry or skill.

Their Orderly Lives

They lived well ordered lives. Theft was unknown amongst them. Polygamy was very rarely practised. Murder was punished by the next of kin, and in the carrying out of this law, there could be no escape on the part of the one upon whom the duty of carrying out the decree fell, if he would preserve the good will of the Great Spirit. True manliness was to be obtained only in war. The warrior who had not won the trophy of an enemy scalp, was but a squaw and not a man. The number of children to each family were few, but sicknesses and pestilences many, so that increase in the number of the tribes was very slow, in many instances at a standstill.

Their Degeneration After the Arrival of Whitemen

The advent of the Whiteman brought nothing of advantage to the Indian. The skins of wild animals obtained in the chase, formerly tanned soft to be made into clothes and conveniences for sleeping,, were, with the advent of the traders, exchanged for clothes, blankets, with cotton prints, usually of gay colors for women's wear. These came into general use and superseded the use of skins for dress. Incentives to industry and skill were taken away; intoxicants were freely distributed amongst them; and the Indians were exploited by all nations alike to help them carry on their numerous wars against one onother.

Friendship with French

The French succeeded in winning the friendship of these four tribes to a remarkable degree. Of course no alienation of their lands to any extent was ever mooted under French rule. The British were different. They entered into the country for agricultural purposes. The transformation of these forests into farmlands meant a decrease in the area of the Indian's hunting grounds. He had therefore to completely change his mode of living, or suffer extermination, in accord with British policy. The French policy, on the other hand,

coincided well with the Indian mode and habit of life. Moreover, the French missionaries, in the main the Jesuit order of
the Catholic church, had been at work very zealously among
these tribes, more especially among the Hurons, and had
succeeded in converting many of them to their faith. This
gave the French an additional hold on their allegiance. Next
in order in their friendship were the Pottawatamies. The
Ojibways and Ottawas stood aloof longest, exerted their
independence more readily, and listened less to white
persuasion than the other two. They were therefore a constant menace to peace in the early history of the province, both
under the French and British regimes. A description of three
of these tribes has been put on record by Isaac Weld as he
saw them in 1796, which it may not be without value for us
to herewith subjoin.

Pottawatamies—"The village of Pottawatamies adjoins the
fort; they lodge partly under Apaquois, which are made of mat-
grass. The women do all this work. The men belonging to
that nation are well clothed, like our domiciled Indians at
Montreal; their entire occupation is hunting and dress; they
make use of a great deal of vermilion, and in winter wear
buffalo robes richly painted, and in summer either blue or
red cloth. They play a good deal at la crosse in summer, twenty
or more on each side. Their bat is a sort of little racket, and
the ball with which they play is made of very heavy wood,
somewhat larger than the balls used at tennis; when playing
they are entirely naked, except a breach of cloth, and
moccasins on their feet. Their body is completely painted with
all sorts of colors. Some, with white clay, trace white lace on
their bodies, as if on the seams of a coat, and at distances it
would be apt to be taken for silver lace. They play very
deep (gross jeu) and often. The bets sometimes amount to
more than eight hundred livres. For playing they set up two
poles and commence the game from the centre; one party
propels the ball from one side and the other from the opposite,
and which ever reaches the ball to the opposite goal, wins.
This is fine recreation and worth seeing. They often play

village aginst village, the Pottawatamies against the Ottawas or the Hurons, and bet heavy stakes. Sometimes Frenchmen join in the game with them. The women cultivate Indian corn, beans, peas, squashes, and melons, which come up very fine. The women and girls dance at night; adorn themselves considerably, grease their hair, put on a white shift, paint their cheeks with vermilion, and wear whatever wampus they possess, and are very tidy in their way. They dance to the sound of the drum and sisiquoi, and the women keep time and do not lose a step; it is very entertaining, and lasts almost the entire night. The old men often dance the Medelinne (Medicine Dance); they resemble a set of demons, and all this takes place during the night. The young men often dance in a circle (le tour) and strike posts; it is then they recount their achievements, and dance, at the same time, the war dance (des decouvertes), and whenever they act thus they are highly ornamented. It is altogether very curious. They often perform these things for tobacco. When they go hunting, which is every fall, they carry their Apaquois with them to hut under at night. Everybody follows, men, women, and children, and winter in the forest and return in the spring."

Hurons—"At that time the Hurons also dwelt on the west side. (The Hurons are also near, perhaps the eighth of a league from the French fort.) This is the most industrious nation that can be seen. They scarcely ever dance, and are always at work; raise a very large amount of Indian corn, peas, beans; some grow wheat. They construct their huts entirely of bark, very strong and solid; very lofty and very long, and arched like arbors. Their fort is strongly encircled with pickets and bastions, well redoubted, and has strong gates. They are the most faithful nation to the French, and the most expert hunters that we have. Their cabins are divided into sleeping compartments, and are very clean. They are the bravest of all the nations and possess considerable talent. They are well clad; some of them wear close overcoats. The men are always hunting, summer and winter, and the women work. When they go hunting in the fall, a goodly number of them remain to guard their fort. The old women, and throughout the winter those women who remain, collect wood in very large quantities. The soil is very fertile; Indian corn grows there to the height of ten to twelve feet. Their fields are very clean, and very extensive; not the smallest weed is to be seen in them."

Ottawas—"The Ottawas are on the opposite of the river, overagainst the French fort; they, likewise, have a picket fort. Their cabins resemble somewhat those of the Hurons. They do not make use of Apaquois except when out hunting; their cabins in this fort are all of bark, but not so clean nor so well made as those of the Hurons. They are as well dressed and very laborious, both in their agriculture and hunting. Their dances, juggleries, and games of ball (la crosse) and of the bowl, are the same as those of the Pottawatamies. Their game of the bowl consists of eight small pebbles (noyaux), which are red or black on one side, and yellow or white on the other; these are tossed up in a bowl, and when he who holds the vessel tosses them and finds seven of the whole eight of the same color he gains, and continues playing as long as he received the same thing. When the result is different the adverse party takes the bowl and plays next, and they risk heavy stakes on all these games. They have likewise the game of the straws, and all the nations gamble in like manner."

The Ojibway or Chippewa Indian Tribes

The Ojibway or Chippewa Indians did not occupy so prominent a place in the history of the Detroit River district as did the other three. Yet these were by far the most important and numerous of all the Canadian Indians. In prehistoric times, they came from the east in company with the Ottawas and Pottawatamies with whom they were joined to form a confederacy known as that of the "Three Fires." They settled first on the east shore of Lake Superior in the neighborhood of Sault Ste. Marie, where they were known, because of the name of the locality, as Saulteaux, and from there bands of them scattered to form villages on the north shore of Lake Superior and throughout Algoma, Thunder Bay and as far west as Winnipeg, Manitoba. In this way they became subdivided into fifty or sixty bands of which the Missisaugas were perhaps the best known.

The Chippewas did not prosecute farming to the extent practiced by the other three nations. Instead they lived chiefly by hunting and fishing. Only a very small number of them moved into the Detroit River district, where they associated themselves with the Ottawas, and took part with Pontiac in his conspiracy against the British in 1763. Subsequent to this event, they continued on terms of friendship with the British, an alliance to which they adhered throughout the War of American Independence and the War of the American Invasion in 1812. In this latter war they were especially helpful in support of the Canadian defence, particularly in the district of Fort Mackinac.

After the close of the last war, no unfriendly action on the part of the Indians against the Canadians has been chronicled in Upper Canada. The Government provided for them reservations where they lived in small bands enjoying annuities from the Government and supporting themselves by farming. In the Detroit River district there are three of these reservations; the one at the Thames, known as the Moraviantown, another at Wampole Island, a third at Sarnia. In Old Ontario, there has been no increase in their numbers on these reservations but in the north they have become a progressive element of the community, chiefly through the activities of missionary agents and industrial schools. Some of them followed farming, others became skilled laborers. They are especially suited to work in the lumber camps and sawmills and many of them make an adequate livelihood in these occupations. Large numbers of them still roam freely and at large in the far north country, devoting themselves to hunting and fishing, an employment made lucrative to them through the activities of the Hudson Bay Fur Company.

POST-OFFICE SANDWICH

The pioneer post-office of the Western District was located at Sandwich, founded as a substitute Capital for Detroit, when that indisputably Canadian Post was ceded to the United States by the British Government, 1796. Erected on Sandwich Street, in the centre of the business section of the town, it is one of the most attractive in appearance of its public buildings.

CHAPTER VII.

Indian Hostility

The Conspiracy of Pontiac

THE reception given to Major Rogers by the French was as peacable and cordial as could be expected from any capitulating and defeated people. The spirit of hostility, secretly nursed by some and openly avowed by others, betrayed them into no rash and unreasoned display of armed resistance. The interchange of garrisons and flags was duly made without any untoward incident happening in respect to any of the posts of this western district. The British took possession; military rule was established; both sides then sat down to await the declarations of the treaty of peace now being prepared between the two nations in Europe.

The Indian Nations Restless

But there was another people, proudly styling themselves a free and independent nation, and the real owners of the land, whose permanent alliance and friendship had not yet been won over on the side of the British. The Ottawa Chief had smoked the pipe of peace with Major Rogers, but openly declared his truce, conditional and temporary. The Hurons went away from his presence giving him freedom to approach the French Commandant unmolested by any ambuscade of their warriors, but their friendly alliance with the French still

remained intact. These both tribes, and with them their
kinsmen, the Ojibways and Pottawatamies, watched curiously
the military procedures carried out between the British and
the French on the morning of the twenty ninth—the lowering
of the Fleur-de-lis and its removal from the flagstaff, the
assembling of the French garrison and the laying down of their
arms, the signing of the pledge of allegiance by the Canadian
militia and the property owners of the colony—all this they
looked upon with wonderment.

This, thought they, served the purpose well as between
the French and the British, but what about them—the true
warriors of the West and the real owners of the land? They
made no concessions; they placed their signatures to no pledge.
Could the transfer of the country be made to any nation without consulting them? If the British soldier thought so, he
was doomed to disappointment. So thought the Ottawa Chief,
Pontiac, during the next two years as he matured his plans to
drive the British out of the land.

The Senecas' Conspiracy

In the meantime, the capitulation of Detroit having been
duly effected, Major Rogers next visited Michilimackinac, received its surrender also, and then returned eastward leaving
Captain Campbell in charge of Fort Detroit. The Captain
found his hold upon the post anything other than secure.
Every new season produced fresh evidence of the hollowness
of Indian friendship towards the British. In the summer of
1761, a deputation of Senecas paid a visit to the village of the
Hurons, the tribe most strongly attached to the French. Their
motive, Captain Campbell ascertained, was for the purpose of
instigating an uprising of all of the Indian tribes, that to-

gether, they might fall upon Fort Detroit, bring about its
fall and the complete annihilation of its garrison. But the
conspiracy, due to the information supplied to the British, was
defeated before it had time to get fully matured. A similar
conspiracy in the summer of 1762 was also nipped in the bud.

A General Massacre of the British Undertaken

In the Spring of 1763, however, there was undertaken
under the leadership of Pontiac, the Ottawa chief, and on a
larger scale than ever before attributed to Indian effort, a
plot to wrest out of their hands, not one, but all of the posts
held by the British in the West, an undertaking which met
with a far greater measure of success than either of the other
two. Pontiac had given his promise that the British would be
permitted to remain in his country as long as they kept up
right relationship with his people, that is to say, as long as
they maintained a regular and generous distribution of gifts
among them. Was it because the British soldiery treated the
Indians too insolently? Or, had the British government with-
held from them an alike measure of gifts and generosities
such as they were accustomed to receive from the French,
and which Pontiac had stipulated would be the price to be
paid for their friendship? Or, were seeds of dissatisfaction
sown through the influence of French habitant and French
trader? Whatever the reasons, whether it was one of these
causes, or a combination of all of them, which induced him to
act, at any rate, Pontiac threw off the mask of friendship
hitherto professed towards the British and came boldly out
with a completed and well-arranged plan to drive them out of
Indian territory. A confederacy of all the Indian tribes was
formed, including the Senecas, the one-time allies of the

British, and plans were matured to fall upon them at every post and settlement at one and the same time, and a general massacre, including every Britisher found in the country, whether soldier, trader or settler, to be effectually carried out.

Pontiac Foiled at Detroit

The day appointed arrived, and the projected massacre was successful everywhere except at Detroit. The garrison of this post consisted of one hundred and twenty three soldiers, with Major Gladwin in command. This post was saved from the fate which befel all the others by reason of the fact that the plan of the conspiracy was revealed to Major Gladwin the day before the time set for its execution. "Concerning the identity of the informant, history has no clue, Major Gladwin very prudently refusing to make the name known."

On the morning of May, the eighth, accompanied by sixty chiefs, representing as many different Indian tribes, all now assembled at Detroit to carry out their bloody purpose, Pontiac appeared at the Fort and asked for a conference with the commandant. Each chief carried a shortened musket under his blanket. At a given signal from Pontiac, each one of the sixty was to despatch a soldier with his musket, and then fall upon the remaining ones with their tomahawks. Instead of taking the garrison by surprise, as expected, he found every soldier in readiness for immediate action. Some stood with drawn swords, others with guns presented, and all keenly alert to the solemn import of the occasion.

Major Gladwin pulled aside the blanket of one of the Indian chiefs and displayed the hidden weapon. Turning to their leader he denounced him for his pretended friendship, his treachery and his murderous intentions. Pontiac answered with unusual dignity and cool bravado. He protested that they came only with a friendly aim. With Major Gladwin's permission, he then turned and passed out, followed by his sixty fellow-conspirators. Neither in his countenance, nor in his actions did he betray any trace of disappointment in the failure of his plot.

His Second Failure Followed by an Indian Siege

A few days afterwards, he appeared again before the Fort, but this time with five hundred followers. Once more he asked for a conference. Major Gladwin offered to admit him, but with only sixty followers as before. This Pontiac refused. Foiled in his second attempt, he now posed in a threatening attitude, proclaiming that he would storm the Fort, starve the garrison, and wrest his country from British occupation and possession. For the next one hundred and fifty three days, he maintained a most persistent effort to carry out his threat.

The defence put forth by Major Gladwin and his little garrison during the adventurous days following this interview is one of the most outstanding and persistent instances of British bravery in colonial history. More than one thousand Indian warriors, well supplied with arms and ammunition, were already assembled at Detroit. One thousand additional warriors from the destroyed forts at Michilimackinac, Sault Ste. Marie and the Ohio river, flushed with success, and inspired to further massacre by the sight of human gore, were on their way to aid their leader to carry out his plot to a successful issue at Detroit also.

Major Gladwin Facing the Danger with Resolution

With courage and resolution, Major Gladwin set his face to the task now forced upon him. In this he showed himself to be an officer of resource and resolution. Were the Indians planning to storm the Fort by a mass attack, or were they planning a siege? In either case, preparations must be made to worthily defend themselves.

He well knew that the number of his men and the stock of his provisions and military equipment were not sufficient to meet the requirements of a long siege. His first undertaking was therefore to despatch a messenger to Niagara in order that reinforcements might be immediately sent him. His next was to make arrangements with the French Canadians whom he could trust to provide him with such provisions as they were able to obtain.

Failure of Relieving Expeditions

On the arrival of the messenger at Niagara, an expedition of ten batteaux set out at once for Detroit. They were not, however, permitted to reach their destination. The Indians were not without information or resource. A war-party, patrolling the shores of Lake Erie, made a successful attack on the convoy at Point Pelee, captured eight of the boats and killed, or made prisoners, all of their soldier crews. The provisions and military equipment, destined for the besieged Fort, were taken to the camp of the Ottawas to replenish the stores of the enemy. A second expedition, comprising 260 men, under command of Captain Dalzell, had a more fortunate beginning, but an almost equally disastrous ending. They reached Detroit in safety, obtained entrance into the Fort, and succeeded in unloading, without molestation, all of the stock of provisions and military equipment which they had with them. Had they been content to remain inactive, until their plans were properly matured, this auxiliary force might have became a real factor in bringing the siege to a successful ending for the British.

A British Sortie Completely Defeated

Instead, they set out early the next morning to make a surprise attack on the Ottawa village. But so far from discovering an unexpecting encampment, they found the Ottawas informed, alert, awaiting in ambush their coming. Some one from the Fort had disclosed their plans. The expedition suffered a complete defeat, and were hastily routed with a loss of fifty nine killed and many more wounded, and but for the bravery of Major Rogers, who was one of the relief auxiliary, and of Captain Grant, who led the retreat, the casualities would have been still greater. This disaster, as well as many others, has been attributed to the influence of French Canadians, secretly hostile, but posing as friends, within the garrison.

Captain Donald Campbell in Treacherous Captivity

One other episode, in which French assistance was rendered to the Indians, remains as a bloody memorial of the conspiracy. On the pretext of a friendly conference, two brave officers, Captain Donald Campbell and Captain McDougall, were decoyed to the house of M'. Beaubien, a Frenchman openly frank in his hatred of the British. There they were treacherously held as prisoners, and eventually put to death in a horribly brutal manner, a commonplace of Indian savagery.

Treaty of Paris Ends the Seige

To those of them who hoped that a reconquest of the country by the French might take place, an incident occurred which had a very depressing effect on their faith. The siege had entered the fifth month of its continuance, when a messenger arrived from the east bearing the news of the passing of the Treaty of Paris, and with him a copy of the terms of surrender of the country to the British. A representative gathering of the French Canadians were assembled at the Fort and the terms of the Treaty read to them by Major Gladwin. This surrender of the French possessions in Canada to the British brought about a complete reversal of the expectations of the habitants of Detroit in respect to aid from the Old Country for the overthrow of the British in Canada which they had hitherto confided to the Indians as a coming event of assured certainty. This changed attitude of mind soon made its impression on the Indians, who were astute enough to realize the significance of the Treaty. Immediately a falling off amongst the supporters of Pontiac began to take place. The first to withdraw were the Hurons who made peace with the British on October 12th, followed immediately afterwards by the Pottawatamies. The Ojibways continued their alliance longer, but at the end of the month Pontiac himself gave up, having achieved a siege, the longest in the annals of Indian warfare. His final submission did not take place until 1766, when a Treaty of Peace was made with him by Sir William Johnson. Three years later he suffered death by murder, while drunk, at the hand of another Indian.

CHAPTER VIII.

The British Become a Divided House

The War of American Independence

Its Effect Upon the Future of Canada and the Detroit River District

FTER British courage had successfully resisted the tactics of Indian strategy, and the loyalty of the French habitants to the terms of the Treaty of Paris had been secured, military rule continued to be administered from Detroit, over the surrounding territory of the West, for twenty five years more, before any serious attempt was made to establish civil rule. The most important event of this period was the revolt against Great Britain of her American colonies. A demand for self-government, in which no element of Old Country influence or authority should find a place, was followed by an uprising in arms, which ended in the establishment of the federal government of the United States of America.

This event, though of great import, was not so far-reaching in its consèquence upon world affairs, as the conquest of Canada from the French in 1763. The establishment of an independent government for the British colonies of North America, was inevitable in some period of their history. No change of destiny was effected from what would with certainty having taking place in some future time in any case. The

method by which it was brought about, civil war, reflected
the belligerent spirit of the age, but no disarrangement of
world affairs was produced by reason of it. The establishment
of successful colonies by Great Britain in America was a
greater achievement than the subsequent granting by her of
political independence to those settlements which she had
established there.

Canada Becomes an Embryonic British Nation

The removal of France from North American affairs was
the first step towards the evolution of Canada as a British
colony, destined to eventually become a new British nation,
sharing with the United States occupation of the continent,
but retaining unbroken continuity with Great Britain. The
establishment of the thirteen British Colonial States into a
new political organization created in North America the same
number of distinctive groups competing for independent
status, as there were preceding the surrender of Canada by
France to Great Britain. The three former, French, British
and Indians, were replaced by, first, Americans, the name by
which the former British colonists, now independent, were to
be designated; second, the Indians, who were still laying claim
to some unmolested portion of their ancestral homeland,
where they could hunt, fish, and erect their wigwams, free
from the menace of interference and alienation of their lands
by white settlers; and, third, Canadians, a new group formed
by the amalgamation of the French, the British fur-traders,
loyalists from the United States, and disbanded British
soldiers, to whom were to be added later, immigrants from
Great Britain and other lands.

What shall become of Canada?

When the Revolutionary war was ended, and negotiations were under way in respect to terms of settlement, the most important question to be settled was, what should be done with Canada? Should it be handed over to the United States, to be incorporated into and become a part of the American Confederacy; or, should it be given an independant status, to work out its own destiny—the evolution of a second great nation on the continent of North America? The answer to these questions was put by the plenipotentiaries of the United States in the form of a demand, that Canada, the whole of it, including Nova Scotia, should be ceded over and become a part of the newly created Republic.

The terms of war settlement were submitted to a commission of whom Dr. Benjamin Franklin represented the United States, and Mr. Richard Oswald, Great Britain. The Commission, while it did not agree to give the whole of Canada to the United States, yet it ceded to them a very important part of the Canadian domain—all of its fur-trading posts on the Great Lakes, all of its territory north of the Ohio, and west to the head-waters of the Mississippi, all of the Detroit river district west, and one half of the lakes and rivers which comprised the drainage system of this Canadian country.

Alienation of Lands arouses Indian Hostility

The first group to feel the effects of these terms of settlement were the Indians. The British house had become divided against itself, but this brought no respite of peace to their nation. It but meant the invasion of their country with a thoroughness inspired by an insatiable demand for more and

more of their territory, to be ended only when there was no
more Indian territory to be acquired. The Indians, therefore,
in their attempts to maintain themselves an independent
nation, occupying such sections of the country as they should
choose to be their own, found themselves at once in conflict
with this newly-created American nation, which demanded of
them the alienation of the great North West territory for
purposes of white settlement.

When the frontiersmen started in to take possession, the
Indians naturally stood in their path, informing them that they
were trespassing on Indian territory, and must therefore keep
on their own side of the Ohio. But, answered these frontiersmen,

"It is ours. We have got it by right of conquest."

"From whom," asked the Indian?

"From the British nation," replied the American.

"Could the British nation give away that which was not
theirs," queried the Indian?

"Well," replied the American, "if we didn't conquer it
from them, we will conquer it from you, now."

And they proceeded to carry out their purpose into effect.

Thus the North West Indian warfare was one of the first
undertakings forced upon the newly-created American nation.

Away back in 1748, there was, as we have observed, a
company chartered to settle a half million of acres of land
south of the Ohio. The response to this call of opportunity
was met by frontiersmen from Maryland and Virginia, whose
presence in the vicinity of the river, the Indian tribes dwelling
north of it, resented. On account of this hostility, the
incoming settlers were not allowed the privilege of a peacable
possession of the land on which they located. They were

constantly being beset upon by roving bands of hostile Indians, hovering around to scalp an unattended or unarmed man and to take away as captives any undefended women or children of the settlement. On the ground that sufficient cause for armed intervention was found in the treatment which was meted out to these frontier settlers, successive armies were sent to the Ohio with instructions to their commanders that the Indians should be cowed to agree to a peacable alienation of all of their lands, both that which was north and that which was south of the river, or, failing to make this surrender they would be compelled into compliance by the arbitrament of the sword, backed by the increasing strength of American arms.

American Indian Warfare affected Canadian interests

The Americans, however, could not dispose of the Indian and his claim upon the Ohio territory, without affecting the interests of Canadians also. At that period of its history, and before the United Empire Loyalists had become a stable part of its population, the settlers of Canada but comprised two groups, the French habitant and the fur-trader. The former had not yet developed any sentiments, other than those of indifference, as to the form of government which should be established in Canada. But the same could not be said of the fur-trader. His spirit was that of restless dissatisfaction with American procedure. The centre of this unrest was at Detroit. This, not only because it was the capital, but chiefly because its trade was in a state of ruin as long as the war between the Americans and the Indians continued. The Indians at war, asked assistance from the fur-trader, and the longer the war continued, the more urgent was their request.

The Leadership of the Fur-traders

The fur-traders were the leaders of the citizenship of the Detroit river district. Their interests were interwoven with those of the Indian nations with whom they were associated in trade relationships. The leaders among these fur-traders were now British-men. Formerly it was Montreal against Albany merchants, but now both were equally interested in the preservation of the Indian fur-trade.

A keen rivalry between these two had formerly existed, a rivalry which dates back to the first attempts of British merchants to get a share of the trade of which the French hitherto had a practical monopoly. In the early stages of the country's history, the great distance between the two colonies and the extent of the unexplored region between them, prevented any clash from arising. As, however, the British frontier moved northward and the French southward, encroachment on each other's territory inevitably occurred. As early as 1740, it is recorded, "over three hundred had crossed the mountains of Pennsylvania with their pack-horses loaded with goods for the Indian trade." Though opposed by the French, they succeeded in establishing a considerable trade with the Ohio and Wabash Indians, and, following up this measure of success, soon began to undermine French influence all along the line, and to establish friendly relations with many nations hitherto trading with the French only. Even from amongst the Hurons, a tribe noted for being particularly friendly to the French, they weaned away a number of them, of whom the tribe of Nicholas, a Huron chief, who lived near the marshes of Sandusky Bay, was an illustrious example.

The Treaty of Paris threw open the whole of the North West to the British for the purposes of trade. Indeed there was given to these a position of advantage over the Montreal merchants. The long distance of that city from Detroit and the Ohio district, the possibilites of innumerable delays in transportation, the frequent handling of the goods on the route, the danger of shipwreck—all these and other reasons, placed the St. Lawrence as the more unfavorable in comparison with the Albany river route for the south-west trade. In consequence, the Albany merchants were in a position to offer better prices to the Indians for their furs, and charge them less for the goods given in exchange, a circumstance which wrought greatly to their advantage. It thus happened, that the fur-trade of North America had passed almost entirely into the hands of British-men when the war of American Independence took place, including those also operating from Montreal.

The Treaty of 1783 established the same difficulties and rivalries as had been in existence when Canada was in the hands of the French. The Americans adopted the same attitude towards their former fellow-countrymen, the British traders in Canada, as they did previously towards the French. They were looked upon by them as hostile rivals, whose domination over the trade of the country must be taken away and the territory placed under American control. The attitude of Americanism to these North West Indians, and the ruination of the fur-trade for which their policies were accountable, produced a strong sentiment, backed by a determined purpose, on the part of the fur-traders, that they should be allowed to shape their own destiny, and build up their own trade enterprises within the domain of Canada, which they were now choosing to be the place of their permanent domicile, free from encroachment by American aggression. The fur-traders,

because the first to evince this distinctively Canadian spirit,
became therefore, the foundation strata of our British
Canadian nationhood.

Two classes of Traders at Detroit

In the Detroit river district there were two classes of
Indian traders, and in any study of that period's history, these
two must be kept distinctly in mind. There was one class,
represented by such men as Robert Dickson, Angus McIntosh,
John B. Askin, and many others. These had no interest in the
war of the American Revolution, as long as it was confined
to the aim for which it was presumably started. It mattered
nothing to them whether or not New York or Maryland
established for themselves a Republican form of Government
as a substitute for British Connection, as long as Canadian or
Indian territory remained unmolested. They were interested
in trade only, not in politics. They, or their parents, had left
the Old Country, some of them Scotland, some England, some
Wales and some Ireland, for the sake of the trade which they
visioned they would establish in Canada. The Revolutionary
War had no interest to them save the influence it had,
detrimental to their trade.

But there was another class, swayed by entirely different
motives and sentiments. They belonged originally, and were
a part of, the United States household. Among these were
several who became leaders in staging opposition to the United
States. In 1777, Alexander McKee, Matthew Elliott, Simon
Girty and four others fled from Pittsburg, and took refuge in
Canada, in consequence of the political troubles rampant in
their own country. Alexander McKee had tasted the
experience of one imprisonment and was threatened with

WILLIAM JOHNSTON McKEE

Great grand-son of the original Alexander McKee, a successful lumberman, of Windsor, Ontario, a former member of the Legislature, who died, 1929, bequeathing an immense fortune to community welfare work in his native county of Essex.

another, he being a member of the Constitutionalist, now the Democratic party, of the United States, when he made good his escape. The Revolutionary struggle was not a matter of indifference to these men. Inspired by party sentiment and embittered feelings, they came to the Detroit district, not to remain inactive, but to augment the forces of their party to the fight. Supported by bands of Indian warriors, they did not feel themselves beaten when the war ended. They did not therefore accept with any spirit of good-will the partition of Canadian territory, and the handing over of the great hunting grounds of the Wabash region to the Republic.

What, therefore should be the reply of these fur-traders to the Indian, when he came, during his conflict with America, seeking their aid?

Neutrality of Fur-trader Impossible

Was it possible for them to maintain an attitude of strict neutrality in respect to the conflict then going on? In all negotiations between whitemen and Indians, the distribution of gifts was the one and only way of maintaining friendship between them. Under the circumstances in which the Canadians now found themselves, the practise of this common custom would require them to do that which the Americans would deem a breach of neutrality. Gun, powder and shot—the ammunition required for hunting purposes—were among the gifts usually given and required. To these would be added provisions. But guns and ammunition, which in the time of peace would be used by the Indian to provide for himself food, could, in the time of conflict, be transformed into munitions and equipment for war. Whether as gifts or in trade, this was the class of goods which the Indians demanded most, during the period of their struggle with the Americans. Compliance with their requests, and comply they must if they would retain their friendship, was construed by the Americans as acts hostile to them, the instigating and urging of the Indians to continued warfare. There was thus being created a set of circumstances, driving the Britisher in Independent America and the Britisher domiciled in Canada, farther apart in national aim and purpose.

Influence of These Traders Under-rated

It will thus be seen that the antagonisms between the two struggling political parties of the thirteen original states of the Union, received opportunity for continuance at the close of the war, but the location of the struggle transferred to Canada. The influence and strength of these emigrated citizens of the United States now domiciled in Canada, were greatly under-rated. Two such men as Alexander McKee and Matthew Elliott, under other circumstances than those then existing, would have exerted very little influence in determining the settlement of so great a question as the international boundary lines separating the United States from Canada, but situated as they were, as part of the Indian Department, first in Pittsburg and then in Canada, they arrogated to themselves places of leadership among the Indian nations, which placed them in control of an Indian army numbering itself in thousands.

But they became an important factor in determining the future destiny of Canada because of another set of circumstances arising from the conditions of the times. The chiefest of these was the universal dissatisfaction with which the treaty of 1783 was received both in the Old Country and in Canada. Why was this?

Dissatisfaction with Treaty of 1783

Any arrangement affecting two nations must be settled on the basis of justice, if a friendly agreement and a satisfied spirit is to be the outcome of it. In the judgment of the British nation and the British people domiciled in Canada, when that treaty was under negotiation, and in the judgment of those who subsequently joined these—Loyalists from the United States and soldiers disbanded from the British regular army—justice in the treaty had been recognized only in its breach. The Americans fought, they claimed, for the principle of self-government. If so, surely this principle should not be entirely ignored, and loyalty to it find no place in this the first international undertaking in which they were called to take a part. The terms of settlement demanded by the representatives of America involved the interests of Canada, as well as those of Great Britain. If Dr. Franklin had not stipulated that the whole of Canada should be voluntarily ceded by Great Britain to the thirteen states of the Union, or if no partition of Canadian territory had been asked for, then it was a matter wholly between the motherland and her revolting colonists; but, since they demanded Canadian territory, surely they would see to it that the people, domiciled in Canada, would be consulted in the matter.

Canada Unrepresented on Commission

"Why" asked Lord Townshend, "should not some man from Canada, well acquainted with the country, have been thought of, for the business which Mr. Oswald was sent to negotiate?"

The question asked so pointedly by Lord Townshend, that question Canadians asked themselves more than once as these negotiations proceeded, with such a man as Mr. Richard Oswald professedly retained to look after their interests. The need of such a representation pushed itself forward more observantly when they reviewed the circumstances under which he received his appointment. According to well-attested history, he was "nominated at the suggestion of Dr. Franklin with whom he was to treat, because he thought he would get along easily with him."*

Dr. Franklin, a man proverbially known as astute, clever, peristent, and in this case at least, grasping, desired that Great Britain should appoint Mr. Oswald as their representative, because he was well suited to espouse the cause of peace, no matter how humiliating would be the sacrifices required of Great Britain in order to obtain it, a peace, the United States in their circumstances at that period could not any longer well afford to do without. Richard Oswald had been a successful Scotch merchant in the city of London; at one time an army contractor, a circumstance which had no doubt something to do in promoting his success. Through his wife, he had acquired large estates in the West Indies and America, and on account of his connection with both countries had occasionally been consulted by the government during the American War. In addition, he had personal qualities which heartily commended him to Dr. Franklin. He was a "pacifical" man, and "a plain and sincere old man who seems

---

*British and American Diplomacy affecting Canada, 1782-1899, by Thomas Hodgins, Q. C.

now to have no desire but that of being useful in doing good."*
This is the man who was to match wits with Dr. Franklin, the representative of the United States, to see to it that justice was done to Canada, as well as to Great Britain.

Mr. Oswald a Weak and Uninformed Representative

His fitness for such an important role is seen in his attitude to the proposal of Dr. Franklin affecting Canada. This, as we have observed, was nothing less that that Great Britain "should voluntarily cede the whole of Canada and Nova Scotia to the United States."*

When this proposal was made, we are told that "Mr. Oswald much liked the idea, and promised that he should endeavour to persuade their doing it."*

When stipulations became somewhat modified, and Dr. Franklin had expressed himself as willing to accept a much more modest cession of Canadian territory, Mr. Oswald again expressed his readiness to serve him. In the discussion of that triangular tract of land between the Mississippi, the Ohio and Lake Erie, a tract out of which there has since been carved five of the greatest and wealthiest states of the Union, a tract which in those early days of its history was designated as "the first spot of earth on the Globe," Mr. Oswald counted it "the backgrounds of Canada, a country worth nothing and of no importance."*

The Fur-traders Resent Their Findings

The Britishers, domiciled in Canada at that time, would have been a supine race of men, indeed, if they accepted with

*British and American Diplomacy affecting Canada, 1782-1899, by Thomas Hodgins, Q. C.

indifference the findings of this Commssion. They were not prepared to accept without protest the carving of the "first spot of earth on the globe," and the handing over of the biggest half to the United States as the price of peace. If that is the price which we must pay for peace, then we are not prepared to pay it. This was the answer of the fur-traders to the proposals of Dr. Franklin. Detroit was then the capital of this "first spot," and the fur-traders were the leaders of the Detroit citizenship of that day, and they were the ones who realized most fully the sacrifice that was being demanded of Great Britain by Dr. Franklin and his colleagues. They had an army of Indians at their disposal. They did not therefore deem themselves without the power of redress. The consciousness of this power added fuel to the flame of their resentment.

The Forfeiting of British Property-rights Also Resented

The dissatisfaction with the terms of the Treaty might have soon blown over, had it not been for the subsequent activities of several States of the Union, in respect to the property rights of American refugee Loyalists, resulting in the formal demand that the claims of these be respected or all of the terms of the treaty be declared abrogated.

Writing to Mr. Adams, Lord Caermarthen said, "You agreed to permit refugee Loyalists to return to the various states to collect their debts. Your states have passed laws prohibiting these people from returning, and confiscated the amounts due them from the states.* You have not kept faith

---

*Mr. John Adams, afterwards President of the United States, was the representative of the Union at London when the Treaty of Peace was being negotiated, and for some time afterwards.

with us, and cannot call upon us to respect a Treaty, when you have not observed it yourself."

The Treaty Stands Eleven Years in Abeyance

For eleven years the Treaty in certain of its enactments remained inoperative, and more than once it was only the commendable self-restraint of the authorities that prevented the re-opening of the war. Fortunately, the desire for armed resistance, which found advocates in both nations, found no countenance from succeeding British Governments. At that particular time, Great Britain was herself in the throes of political unrest. Opposition to monarchical government was sweeping over Europe with increasing popularity, the "United Irishmen" were stirring up unrest in Ireland, and East India was a boiling pot of sedition. Besides her people were tired of war, and most of all, a war against her own flesh and blood. This is probably the explanation for their acceptance of a treaty which never did, and which never could, win the respect of her subjects.

"It has been frequently said," wrote a worthy authority, "that of all the treaties executed by Great Britain, this Treaty was one in which she gave most and took least . . . . . . her surrender presents an instance of apparent sacrifice of territory, of authority, of sovereignty and of political prestige, which is unparalleled in the history of Diplomacy."*

It was not until one of America's greatest citizens, Chief Justice John Jay, had staked his political reputation on the side of a peaceful settlement, that a second treaty was enacted which removed the boundary question for a time from the arena of controversy, and secured for the country an era of

---

*Wharton, in his History and Digest of International Law.

sixteen years of peace.

In one respect in which the War of the American Independence affected profoundly the destiny of eastern Upper Canada, it had little, if any, influence on the Detroit district. Settlement of Upper Canada first began, as has been observed, from the Detroit River district eastward. A subsequent settlement began from the east, westward, made up of Loyalists, American citizens who were forced to emigrate from the United States because of the treatment accorded them at the hands of the successful Revolutionists. Of this class of settlers, few, if any, came into the province by way of Detroit. Apart from the disbanded soldiers of the British Regular Army and some of the Hessian troop, the settlers who came into the district from the United States immediately following the close of the hostilities, were German families who came to Canada from Europe by way of Pennsylvania. During the passing of the Jay Treaty, many of these settled in the State of Michigan where they were known as 'Dutch Tories'. A number of them crossed the river and became a part of the Lake Erie and Thames River settlements. The prospect of obtaining good land, this and no other, was the motive inducing them to migrate to Upper Canada. The term, Loyalist, applied to either of these two classes of settlers was an unwarranted use of the title. Of genuine Loyalists, they treked into Upper Canada from the east, westward, but no colony of them nor any appreciable number of them, found their way into Detroit. The fur-trader and the French habitant were the dominant settlers of the west until the coming of migrants from the British Isles in numbers, gave to the population its present-day complexion.

CHAPTER IX.

First Representative of Civil Law and Order

The Detroit District is Named Hesse

The First Court of Justice within its Bounds, Established by Proclamation of Lord Dorchester, 1788

IN the interval, pending settlement, and notwithstanding the chaotic condition of affairs in the west, Lord Dorchester, the Governor of Canada, found himself compelled to take a forward step in the administration of the government of the province placed under his rule. In 1788, he made proclamation that the country west of the Ottawa river be divided into four judicial districts, which he named respectively, Lunenburg, Mecklenburg, Nassau and Hesse.

Lunenburgh comprised the easterly area, the western boundary of which was the Gananoque river. Macklenburg comprised the territory between this river and the Trent. Nassau extended westward from the Trent river to a meridian line passing northward from Long Point on Lake Erie. "The District of Hesse," the proclamation stated, "is to comprehend all the residue of our said province in the western or inland parts thereof, from the southerly to the northerly boundary of the same." . . .

The territory of Hesse comprised, therefore, all the area in the subsequent province of Ontario drained by the Thames river and northward to the Hudson Bay territory. It extended westward to include, in addition, all the territory north of the Ohio river and eastward from the headwaters of the Mississippi, to the Detroit river, that section of Canada which the Oswald and Franklin Commission had agreed to hand over to the newly-created republic of America.

This division of the province into districts was made for judicial and land settlement purposes. Hitherto, in civil matters, the Detroit river district was a law unto itself. The Law Courts were held at Montreal, and that was a very long way off, hence little use was made of them except in cases of extreme emergency.

Districts formed for Civic Purposes

The only departures from law and order which they seriously recognized were unpaid debts, theft and murder, but the instances of any of these were very rare. In respect to unpaid debts, the traders had the remedy in their own hands. No beaver skins meant no more ammunition, blankets, flour or pork. Necessity, therefore, compelled the debtor to produce the beaver skins. In some instances, in payment, the man himself or some member of his family would be bound over to service for a number of years, a species of slavery. The punishments meted out for crime reflected the spirit of the age. For theft, a man was whipped, or hanged. For manslaughter, his hands were burnt. For wilful murder, he was put to death. In the main, all authority was focussed in the military commandant, both in civil and military matters.

The French Canadians lived a very simple life. They had an unwritten code for right procedure among themselves, and all of them, with rare exceptions, kept themselves within the bounds of their self-created order. But with the incoming of British soldiers and settlers in increasing numbers, taking up their permanent abode in the district, there was need of the restraining hand of judicial authority and the appointment of a Board for the proper administration of the sale and grants of land for settlement. This need, Lord Dorchester supplied in 1788, when he called into existence a Law Court for the Detroit district.

The First Court Officials

In the first draft of its organization, the judicial system which Lord Dorchester set up was made to consist of three justices, a sheriff, and a clerk of the court, which was known as the Court of Common Pleas. There were also eight magistrates appointed and a coroner. The magistrates appointed were Alexander Grant, William McComb, William Caldwell, Matthew Elliott, Guillaume LaMotte, St. Martin Adhemar, Joncaire de Chabert and Alexander Maisonville. Gregor McGregor was appointed sheriff, Thomas Smith, clerk of the Court, and George Meldrum, coroner.

Among the appointees, it will be noticed that there were five bearing French names. This was in accord with the well-advised policy of the British, to give equal recognition to the French in all matters affecting the government of the country. From the very first, the aim was to give equal standing to all peoples domiciled in Canada, placing no disability on any one, or lessening his privileges of citizenship because of the accident of his race or religion.

The three justices appointed—Duperon Baby, Alexander McKee and William Robertson—declined to act on the ground that they were but ordinary laymen, while the appointment required a man qualified by training and experience for the position. A Petition was accordingly forwarded to Lord Dorchester, requesting that such an appointment be made. Lord Dorchester admitted the wisdom of their petition, and had William Dummer Powell, an able lawyer of Montreal, appointed the following year.

William Dummer Powell becomes First Judge of Ontario

William Dummer Powell, the appointee, and therefore the pioneer Judge of Upper Canada, descended from a prominent family of Wales, whose villa was known by the name of Caer Howell, Montgomeryshire. He represents the third generation of his family domiciled in America. His grandfather John Powell came out in the early days of its colonial history, and settled in Boston, where he married Anne Dummer, a sister of Governor Dummer of Massachusetts. The original Dummer, a Roundhead and Independent, emigrated from England at the time of the Stuart Restoration in 1660. The eldest son of John Powell and Anne Dummer, also called John, became a prominent merchant of Boston, married Janet Grant, a daughter of Sir Sweton Grant, a wealthy merchant of Rhode Island, a lady distinguished both for her personal beauty and literary acquirements. William Dummer was the oldest of their five children. He was born in Massachusetts, November the fifth, 1755, and lived through the troublous days of three wars, his own life, under the fate of circumstances, as troubled as the era in which he lived. He was intended by his family for a mercantile life, but drifted instead, into the legal profession.

William Dummer Powell received a thorough, preparatory educational training for his future life. After the first years of tuition under his mother, he was sent to the free Grammar school at Boston, where he spent three years. Thereafter he was sent to a private school in England, where he remained until he was fourteen years of age. This was followed by a two year course in Holland, where he was sent to learn the Dutch and French languages. When he came back to Boston at seventeen years of age, his acquirement of a thorough knowledge of the French language, both colloquial and literary, was perhaps not the least valuable of his attainments.

Shortly after his return to America, he experienced the first of the many disappointments which were to dog the path of his future life. The summers of 1773, and 1774, he spent in Montreal and Canada, but in the winters devoted himself to the study of law, under the Attorney General of Massachusetts, not to fit himself for the practise of law but to better prepare for a business career. In the autumn of 1774, his mother was accompanying him to New York, to use her influence to find a place for him in business among friends in that city, but on their way there was stricken with an attack of small-pox, from which she did not recover. Her son alone was present with her at her death "and friends did not even assist in her interment for fear of infection."*

The Troubled Political atmosphere of America

By this time, the political atmosphere in America was becoming increasingly troubled, and the time was rapidly approaching when every citizen had to declare himself on one

---

*Her son was very proud of her, and fifty years after her death spoke thus of her: 'My mother is still remembered for her charms of person and mind. She was the most perfect beauty of the burnette character I ever saw, and at the age of 39, when she died, though somewhat corpulent, was constantly taken for my sister, as we travelled together from Boston to New York where she died of small pox in the year 1774'."—(Life of William Dummer Powell: Riddell).

side or the other, William Dummer Powell, following the example of his father, took the side of the Constitutionalists, though two of his father's brothers took the side of the Revolutionary party. Discouraged by the disagreeable state of affairs in his native State, the senior Mr. Powell purchased property in Shropshire, England, and moved there with his family in the autumn of 1776. This step brought down upon him the wrath of the insurgent authorities, who declared his properties forfeited, and his return to the State forbidden.

His Marriage

The year after his mother's death, William Dummer married Anne Murray, a daughter of Dr. Murray, of Norwich, England, while she was on a visit to Boston, and that same year returned with her to England. While there he continued his study of law, attending faithfully the sittings of the Courts at Westminster Hall, London. Although he had completed the required course of study to enable him to be called to the Temple Bar, lack of funds prevented him from adding this honor to his other qualifications for the practise of his profession until 1783. In the meantime he decided he would emigrate to Canada, and begin there the practise of law.

Opens a Law-office in Montreal, 1779

Fortified with letters of introduction from the Secretary of State to Sir Frederick Haldimand, Governor of Canada, and other important personages, he set sail in the midsummer month of 1779 for Quebec, leaving to his great regret, his wife and three little boys behind at her father's home. On account of his being a native-born American, his reception was not, on the part of Governor Haldimand, as enthusiastic as he anticipated, hence he deemed it wiser to set up practise at Montreal, where he had several friends and acquaintances. The first client who sought his services was a wealthy Hougenot, who had earned the antipathy of the Montreal Courts by his free speaking, and from whom therefore he claimed he could get no fair dealing. Mr. Powell had already been apprised of the likelihood that this man would solicit his aid, and had been warned by no less a person than the Attorney General, Mr. Monk, to accept from him no retainer on account of his ill-standing with the courts. Mr. Powell answered this test of his courage and the integrity of his moral character, by accepting the retainer, and with a special jury made up principally of English speaking citizens, he succeeded in winning a verdict in favour of his client. This favorable beginning raised him to the front rank among the lawyers of the country. But it won for him also, the hostility of prominent men in the profession and higher circles of Montreal society. In the midst of the violent rivalries of races and religions, so conspicuous in those early days of Canada, the person who would seek to maintain a firm loyalty to principles of rectitude, was not preparing for himself a bed of roses. This William Dummer Powell did not take long to discover.

Agitates for Revision of Quebec Act

The first important political undertaking with which he became identified was in connection with an agitation for the revision of the Quebec Act. This enactment was made to please the French, but its terms, though tolerated, were never acceptable to the British. The need for its revision was emphasized at that particular time, as many Loyalists were

moving from the revolutionary zone, and were hindered from
making Canada their place of selection because of their dislike
to the French laws in which trial by jury and the Habeas
Corpus, time-honored features of the British law, found no
place. A petition from the English speaking British and
Americans in Canada to the King and Parliament of Great
Britain was formulated, 1783, and a delegation consisting of
Mr. Powell and two others were appointed to support it before
the appointed authorities in Great Britain. The petition was
in the main successful—recognition of the Habeas Corpus
was authorized April 29th, 1784, and trial by jury in 1785—
and the influence of the petition was without doubt one of the
inspiring causes leading to the passing of the Constitutional
Act of 1791, which gave to Upper Canada British laws in both
civil and criminal jurisdiction, and to Quebec, British laws for
criminal, though retaining French laws in respect to civil
matters. The achievement of these changes from the provisions of the Quebec Act opened the door for the settlement
of Loyalists in Canada. Had it not been for this concession to
British sentiment, few, if any, of these would have chosen this
country for their future home, and since Mr. Powell was the
inspiring influence behind the petition, he merits a just
appreciation for the part he played in bringing it about.

Friendship of Lord Dorchester

On his return from this visit to England, followed by a
short period spent in Massachusetts, he found Sir Frederick
Haldimand removed and Lord Dorchester returned to occupy
his former place, as head of the Government of Canada. He
was favorably received by the new Governor, who honored him
with considerable government patronage. When the petition

WILLIAM DUMMER POWELL
(1755-1834)

THE FIRST JUDGE OF ONTARIO

Appointed in March, 1789, by Lord Dorchester, to preside over his newly-created court of common pleas for the district of Hesse. He held his first court at Assumption, now Sandwich, July 16th, 1789, and continued to preside over it until it was abolished in 1794, when he removed to Toronto and became a judge of the court of King's Bench for Ontario which superseded it.

from Detroit arrived asking for the appointment of a trained official, instead of the three lay justices who had been appointed, Lord Dorchester offered him the position. At that time, he enjoyed the distinction of being the foremost lawyer in Montreal, yet in the face of the bright prospects before him in the practise of law, he accepted the Detroit post, with the expectation, it would seem that when the organization of the judiciary on the basis of British law and court procedure would take place he would be appointed its head. He moved with his family in the midsummer month of that year, to Detroit and held his first court at Assumption, now Sandwich, July 16th, 1789. This is therefore the beginning date for the establishment of civil courts in Upper Canada, although as yet it was not a separately organized province but still a part of the province of Quebec.

Difficulties to Maintain Strict Integrity in the West

If the Scriptural statement that a man is born to trouble as the sparks fly upward needed corroboration, the life of Upper Canada's first Judge would supply ample evidence of its truth. The Detroit river district at that time was filled with mercantile adventurers living on trade with the Indians, a trade which it is said totalled a sum of between $750,000 and $1,000,000 annually, throughout the province of Canada. Some of these men may have been honest merchants but in the main, inspired by avarice, they stooped to intrigue and all manner of dishonesties to gain their ends. To expect these to tolerate the activities of a native-born American, who was sent there for the express purpose of establishing the practise among them of the principles of honesty, integrity and rectitude, would be to court disappoint-

ment. The spirit of the age is well exemplified in respect to
two cases which came up for trial at the first criminal court
held in Assumption after the organization of Upper Canada
into a separate province, the onee involving theft of property
the other homicide. The first, a negro, charged with theft of
property from a Frenchman of Detroit, was tried, found guilty,
and sentenced to be hanged, which was duly carried out. The
other case involved several men charged with brutal assault
upon an Indian at Michilimackinac, followed by fatal results.
Though the persons implicated in the crime were known, yet
supported by the backing of the community, they were not
subjected to the inconvenience of an arrest nor asked to put
in an appearance before a jury. In an age which deemed theft
of property a crime punishable by death, and murder a
convenient accident or necessity, a representative of law and
order who regarded sacredly the duties of his office, need not
expect to escape that disfavor inevitably following the exercise
of the traits of firmness, courage and loyalty to principle,
which Judge Powell exemplified. It was not long, therefore,
before he found himself attacked by those whose interests
were not promoted by honesty and fair dealing, especially in
connection with transactions affecting the purchase and sale of
Crown Lands. He was a member of the newly-created Land
Board, to whose opinions all the other members paid a marked
and respectful deference. He was faithful in attendance on its
meetings, so that few of its transactions escaped his knowledge
and supervision. He was especially attentive to see that the
Indians should not be imposed upon by unscrupulous persons,
practising the dishonest exchange of a large tract of land for
a trifling price, the generous distribution of a few dollars worth
of intoxicants oftentimes putting a man in possession of

thousands of acres of valuable property. Both as a judge, and as a member of the Land Board, he refused to recognise such transactions as valid, and interposed himself in opposition to their subsequent ratification by the Government. This brought him into disfavor with the thirsty Indians who wanted to sell, and the speculative adventurers who wished to make the purchase.

Plots to Ruin the Judge

These worthy endeavours to bring the community life of the district into subjection to a well-arranged system of law and order, gave rise to two dastardly plots, the one against his reputation, the other against his life. The latter, the plot of an Indian chief, was easily circumvented through the timely warning given Judge Powell of his intended purpose.

But the second, the attempt to ruin his reputation and future on a charge of treasonable disloyalty was foiled only in a measure. A letter, a clever imitation of his handwriting, purporting to be written to Secretary Knox of the United States but unsigned, was placed in his office, and when asked if it was his handwriting, on a cursory glance, Mr. Powell answered in the affirmative, but on ascertaining its contents declared it a forgery. Had this letter stood alone, its influence would have been soon counteracted, for not only was it proven a forgery, but the persons who were party to it, were discovered. Unfortunately there was added to its influence the accident of his birth-place. It was only necessary for his enemy to spit out, "He's an American," and this would be enough to surround his name and reputation with a prejudice which could not be other than an obstacle to his influence and a hindrance to his promotion. One would hesitate to associate the name of so distinguished a person as Governor Simcoe with so puerile a prejudice, yet when the appointment of a Chief Justice was to be made in 1792, a position to which Mr. Powell, above all others, was justly entitled, his name was passed over; and when a second time, in 1794, the position was again vacant, notwithstanding that he knew that the claims of Judge Powell were supported by the home authorities, he wrote back to England and said, we want an "English" lawyer for the position. It was not until his valued services during the war of the American Invasion, and the friendship of Sir Isaac Brock and Lord Prevost and their faith in his abilities, had rid the public mind of all remaining vestiges of suspicious prejudice, that his services were accorded public recognition.

Author of Brock's Proclamation

With the probable exception perhaps of Governor Simcoe, Judge Powell was highly esteemed by all of the Governors of the province, but it was to Governor Gore that he owed a continuance of the favors granted to him by Lord Dorchester. A mutual confidence and friendship existed between Sir Isaac Brock and the Judge, which no doubt would have found tangible expression if the General's tragic death at Queenston Heights had not taken place. However, to this friendship, we owe a worthy product of Judge Powell's wisdom and prudence, in that famous historic document, the proclamation issued by Sir Isaac Brock in answer to General Hull's, in the midsummer month of 1812, on the occasion of the outbreak of war between the United States and Great Britain. In this proclamation, an imperishable heirloom unto us from the past, he became the spokesman of Canada, and with dignity, yet unyielding firmness, he sets before the world, the political aims of the embryonic nation. This document, and its production, expressing as it does Canadian faith in British institutions, and Canadian purpose to loyally defend them, will doubtless constitute the most enduring monument of his life's achievements. It loses nothing in prestige, though in his own life time, his loyalty was so oftentimes questioned and challenged by reason of the accident of his birthplace. Indebted as the country undoubtedly was to his eminent services in its behalf, it is a matter of satisfaction to succeeding generations, that recognition, though belated, was made and that he received from Governor Gore the promotion that ought to have been his a quarter of a century earlier.

Subsequent Offices and Emoluments

He was appointed a Legislative Councillor, a member of the Executive, and later speaker of the Legislative Council. In 1816, he was appointed a Chief Justice, this, said to be, the crowning ambition of his life. In the exercise of these offices, he became one of the most prominent men of his day. His services in these spheres brought him both evil report and good report. He was counted a member of the Family Compact, and did not escape the odium attached to the name. With the approach of age, his abilities and influence both suffered decline. He retired at seventy on an annual pension of £5000, which he lived for nine years to enjoy.

Domestic Anxieties and Troubles

In his domestic life, he did not escape the vicissitudes of care and sorrow as well as of joy inevitably associated with

the rearing of a numerous family. Of his eight children, one
died in infancy, two sons died in young manhood, three married
and left issue, some of whom were represented among the
most prominent of Toronto's families. On the part of two of
his children, there was forced into the arena of his personal
life the things which produce anguish to the parental heart.
His youngest daughter, Anne, figured in an infatuation which
brought humiliation into her father's household and tragedy
into her own life. One of his sons, a successful merchant in
New York, barely escaped an ingominious death. Inveigled by
a political adventurer, into an effort to secure independence
for the island of St. Domingo, he was arrested by the Spaniards,
saved from subsequent punishment by the appeal of his father
appearing in person in his behalf before the Government of
Spain, an incident which indicates the influence and energy
which characterized his whole life.

Death and Burial, Aged 79

He spent the declining years of his life in Toronto, where
he died September 6th, 1834, and was buried first in his own
burying ground, but later removed to St. James' cemetery,
Toronto.

As the pioneer and first Judge of Ontario, he gave to the
office which he held, a prestige which can only come from the
tactful exercise of the qualities of firmness, honor, courage,
and moral rectitude, and by so doing, he laid the foundations
on which there has since been built up that respect and high
regard which our people have toward our courts of justice and
their officials.

CHAPTER X.

The Western District Organized

The Pioneer Effort to Establish Municipal Government in Upper Canada, by the Organization of Districts and Counties

THE name Hesse, as representing the Western section of Upper Canada, is associated with the four years' civil rule of this province under Lord Dorchester preceding the passing of the Constitutional Act in so far as there was any civil administration established. The name Western district, introduces us to Colonel James G. Simcoe, his governorship, his plans and policies for Upper Canada. It was one of his first acts to give the District this name and hence the name stands pre-eminently associated with his rule.

The Boundaries of the Province Undetermined

When he came into office, Detroit was the capital and at that time the centre of the Western District. During his whole regime the boundaries of the province were in a state of unsettlement. While the Treaty of 1783 and the commission appointed to effect a settlement had come to an agreement in respect to the boundaries between United States and Canada, this had not been finally ratified until the passing of the Jay Treaty in 1794, which gave two years for final ratification. In addition, the northern and western sections of the province

119

were a great and unexplored region of which no definite
knowledge could be had until further exploration took place.
Governor Simcoe found himself, therefore, in the same position
as Lord Dorchester, with a province over which he was to
establish a rule and yet the bounds of that province not
determined.

In furthering the establishment of a civil administration
over the province, one of his first acts was to divide the
province in so far as it was known and explored, into nineteen
counties. Lord Dorchester had divided the province into four
districts, as the first step in establishing civil administration,
but to these he gave German names.* Governor Simcoe
changed these to English names. To the fourth, Hesse, he gave
the name "The Western."

The Establishment of Counties

The division of the province into Counties could not at
first be anything other than preliminary. Until the boundaries
of the province were definitely and permanently fixed, and the
territory further explored, map making was mainly a
conjectural undertaking. A rough map at the best only could
be made. The southern section of the province, as far east as
the Ottawa river and as far west as the Detroit, had been well
explored. The lakes and rivers which formed its southern
boundary, had been navigated. Seamen had patrolled the
whole shore line. This boundary was, therefore, sufficiently

---

*An Act was passed in the year 1792, for building a Court House and gaol
in each district within the province and changing the names of the said
districts:
  Lunenburgh—He named Eastern;
  Mechlenburgh—Midland;
  Nassau—Home;
  Hesse—Western.

explored to give a general idea of the shore line. Working from this as a basis, he divided the whole southern frontier into a series of counties, their southern boundaries definitely fixed, and their eastern and western boundaries, the meridian lines, running north and south from their south-east and south-west extremities, leaving their northern boundaries quite indefinite.

By Proclamation, dated July 9th, 1792, these counties were named as follows: Kent, Essex, Suffolk, Norfolk, Lincoln, York, Durham, Northumberland, Hastings, Prince Edward, Lennox, Addington, Ontario, Frontenac, Leeds, Grenville, Dundas, Stormont, Glengarry.

The first eight of these are the names of the eight eastern counties of England. The townships making up these eight counties were named after important towns in the similar counties of England.

Kent County as the most westerly of all these nineteen counties, included Detroit and Michigan. It was the only one of all the counties whose future destiny was affected by the dispute in respect to the boundary between the States of the Union and Canada. It included everything that was left over of the other eighteen counties, and hence all of the disputed territory was within its bounds.

County Lieutenants Appointed for Civil and Military Purposes

Governor Simcoe, as did Lord Dorchester, found himself occupying an office that carried with it twin responsibilities —the one civil, the other military. The division of the Province into counties was with a view to carrying out more efficiently his administration of both departments.

Over every county there was appointed a County Lieutenant an office which corresponded to that of the Lords Lieutenant of England and Ireland. This also was done for a twofold purpose. Like that of the Governor, the County Lieutenant was both a civil and military official. As a civil administrator there was laid upon him a certain responsibility in the granting of land to settlers and seeing that justice was done to them. He possessed also the right of appointing magistrates. The second responsibility, the military, was the more important of the two. It was his duty to keep an enrolment of men eligible for military service, and to see that they were kept in constant readiness for service. With the ownership of the county in dispute, with belligerent armies hovering on the frontier and demonstrating themselves ready for hostile engagements, with Indians as neighbours, who to-day might be friendly but to-morrow an enemy, the settlers, as well as the administrators set over them, found themselves compelled to keep themselves in constant readiness for war. The enlistment of men for service, the calling of them into action and the provisioning of them when they were on duty, the responsibility of these things, in his own county, rested upon the Lieutenant. Governor Simcoe appointed Alexander Grant, Lieutenant for Essex and James Baby for Kent. The office of County Lieutenant continued until 1849, when the present municipal system was established.

A second reason for the division of the province into counties was for the purpose of facilitating the election of members to Parliament. There was no local government of these counties established before 1850. There was in fact no need for such government, because of the sparseness of settlements and the few regulations required to maintain law and order in the county. In some counties as in Suffolk, there were no settlers whatever at the time when the first parliament assembled at Newark. The Legislative Assembly and the

Legislative Council, with the Executive, looked after all local needs, which could not be taken care of by the local Lieutenants and the magistrates under them.

The First Boundaries of the Western District

The formation of the Province into districts was for the purpose of making administration of justice more easy. In the first instance, the Western District was made to consist of the counties of Essex, Kent and Suffolk, the seventeenth, eighteenth and nineteenth counties of the province. The boundaries of those were designated as follows:

"That the seventeenth of the said counties be hereafter called by the name of the County of Suffolk; which county is to be bounded on the east by the County of Norfolk, on the south by Lake Erie, until it meets the carrying place from Point au Pins unto the Thames, on the west by the said carrying place, thence up the said river Thames until it meets the northwesternmost boundary of the County of Norfolk.

"That the eighteenth of the said counties be hereafter called by the name of the County of Essex; which county is to be bounded on the east by the County of Suffolk, on the south by Lake Erie, on the west by the river Detroit to Maisonville's mill, from thence by a line running parallel to the river Detroit and lake St. Clair, at the distance of four miles, until it meets the river La Tranche or Thames, thence up the said river to the northwest boundary of the County of Suffolk."

"That the nineteenth of the said counties be hereafter called by the name of the County of Kent; which county is to comprehend all the country not being territories of the Indians, not already included in the several counties hereinbefore described, extending northward to the boundary line of Hudson's Bay, including all the territory to the westward and southward of the said line, to the utmost extent of the county commonly called or known by the name of Canada."

The Second Boundaries of the Western District

The boundaries of the Western District were fixed a second time by Statute passed in 1818. This made the district to consist of the counties of Essex and Kent and the inhabited part of what later became the County of Lambton together with all north of these two counties which was not included in the district of London. There is a map drawn by David W. Smith which sets forth the limits of this Western District as well as that of all the other districts of the province. This Act of 1818, organizing this section of the province into a District reads as follows:

Upper Canada Statues Revised—1818

"And be it further enacted by the authority aforesaid, That the Township of Dover, Chatham, Camden, distinguished by being called Camden West, the Moravian tract of land, called Orford, distinguished by Orford North and South. Howard, Harwich, Raleigh, Romney, Tilbury, divided into East and West with the township on the river Sinclair, occupied by the Shawnee Indians, together with the Islands opposite thereto, do constitute and form the County of Kent."

—Section 38

"And be it further enacted by the authority aforesaid, That the townships of Rochester, Mersea, Gosfield, Maidstone, Sandwich, Colchester, Malden, and the tracts of land occupied by the Huron and other Indians, upon the strait, together with such of the islands as are in Lakes Erie and Sinclair, or the straits, do constitute and form the County of Essex.

—*Section 39*

"And be it further enacted by the authority aforesaid, That the counties of Essex and Kent, together with so much of this province as is not included within any other District thereof, do constitute and form the Western District.

—*Section 4o*

The Third Boundaries of the Western District

Again in 1845, the boundaries of the Western District received a further revision. The County of Huron was now being organized and the boundaries between it and Kent County had to be indicated. The Western District was, by Statute passed that year, made to consist of the two Counties of Essex and Kent only. The townships in each of these two counties were set forth, eight in Essex and twenty-one in Kent. Those in Essex were, Anderdon, Malden, Rochester, Gosfield, Maidstone, Mersea, Colchester, and Sandwich. The twenty-one townships of Kent, listed alphabetically were: Bosanquet, Brook, Camden, Chatham, Dawn, East Dover, West Dover, Enniskillen, Howard, Harwich, Moore, Orford, Plympton, Raleigh, Romney, Sarnia, Sombra, East Tilbury, West Tilbury, Warwick and Zone.

District of Kent Formed, 1847

It will be seen from the above list, that all of the ten townships, which later became the County of Lambton, are at that date, included in the County of Kent. Two years later, 1847 (under the guidance of Joseph Woods, the member for

Kent) the Western District was divided into two districts, all of the twenty-one townships except Tilbury West were made into a separate district called the District of Kent. West Tilbury was joined to Essex County, and with this addition to it, and Sandwich as the judicial seat, it formed the continuance of the Western District.

Very minute regulations were passed by this Act, in respect to the organization of this new "District of Kent." It was stipulated that a gaol and Court House be built at Chatham, on the site which had been previously reserved for them in the survey of the town plot, which was made the District capital or seat of the District government. Regulations were made for the election of District representatives, for borrowing money to build this Court House and gaol, for the appointment of a Building Committee and many other regulations as a guide in the formation of the District.

Districts Abolished, 1849

But this organization was not long-lived. Two years after, 1849, was passed the Municipal Act which organized the province into Counties and established local government for a group of townships, which were made to form the County system as we have it to-day. The ten townships of the north were separated from Kent and made into a new County, to which was given the name, Lambton, commemorative of the great statesman who was honoured by the home government to bring about that beneficial union of 1840, an enactment which is rightly described as the greatest of all progressive enactments made for the benefit of the province.

The Municipal Act of 1849 came into force the following

year. From this time forth the name Western District, passed out of use. As the county increased in population, the local governing bodies increased in number, with a corresponding decrease in the size of the territory which was included under the supervision of the District or County Council.

Peculiar Feature of County Boundary Line

In the history of boundary lines, Essex suffered very little change from what it was when first delimited by Governor Simcoe. A strip four miles wide running south from the Thames river and parallel with the Detroit was made a part of the county of Kent. As the counties were established to facilitate the election of members of parliament for the newly created House of Assembly, this strip of land was taken off Essex in Order that the Parish of Assumption might be included in the county of Kent, together with the town of Detroit and the French colonies surrounding it. Detroit was of course, at this period according to Treaty ceded to the United States, but as its inhabitants were British subjects they were reckoned as a part of the Province of Upper Canada and the County of Kent until the evacuation of the district lying on the west side of the river should officially take place. The anomaly of people dwelling in what was considered a part of the United States, taking part in the Government of Upper Canada is an evidence of the fluid nature of the political conditions of the time. Though the territory had been ceded the people themselves were still British and would continue to be until they took the oath of allegiance to the United States, or moved over to Canada.

The greater number of the colonists were with this

arrangement of the county boundaries domiciled in the County of Kent. It included all the settlers north of the Thames River, all in the parish of Assumption on the left bank of the river, and all those in the State of Michigan. The County of Essex contained the people dwelling in Petite Cote the New Settlement at the mouth of the Detroit River and on the north shore of Lake Erie and all on the south side of the River Thames. In the distribution of the members of parliament, Kent was given a representation of two, and Essex one, because of this inequality in their numbers.

In a later period in the history of the Western District, a District Council was established. Its first meeting was held in 1492 and representatives were appointed from all the organized townships then established in the district. This Council continued to function until 1847 when, as has been observed, it was subdivided, and two District Councils were formed in 1847 to be followed by the abolition of districts in 1849.

The Electorate of Bothwell Organized, 1877

For electoral purposes, another division was made. The electoral constituency of Kent, for parliamentary purposes was made to consist of the townships of Dover, Tilbury East, Romney, Raleigh and Harwich and the town of Chatham.

Another electoral division was formed called Bothwell, which comprised seven townships, three from Lambton, Sombra, Dawn and Euphemia; and four from Kent—Zone, Camden and the Gore thereof, Oxford and Howard. This continued for years, after which the name Bothwell disappeared from off the map, and the electoral divisions as we now have them came into vogue.

ASSUMPTION COLLEGE, SANDWICH

Founded in 1855 by the Jesuit order, taken over by the Basilians in 1870 with Father O'Connor as superior, and now affiliated with the Western University, London, it promises to become, if progress continues the pioneer university of Essex County. Three hundred and fifty students were enrolled in 1925, the fiftieth year of its history under Basilian auspices.

CHAPTER XI.

The Year of Peace

The Jay Treaty and the Exodus of 1796

THE year 1794 goes down in the annals of history as a crucial year in the affairs of the Detroit river district, and this because of both what happened and what also was on the verge of happening, a second war between Great Britain and the United States. There comes a time when a problem producing disagreement between two nations can not any longer await solution, a settlement between them any longer be delayed. For eleven years, the evacuation of the outposts, in accord with the terms of the treaty of 1783, between Great Britain and America stood in abeyance. The people inhabiting the ceded territory were still British subjects, and would continue as such until they took the oath of allegiance to the United States. The situation created was abnormal; in some instances intolerable, the Civil Court established by Lord Dorchester was for the sake of convenience and increased efficiency, but in what place should it be held. With the arrival of William Dummer Powell to become the first Judge of the district, although he had his residence in Detroit, the sittings of the Court were necessarily held at Assumption—since become the town of Sandwich—this in acknowledgment of the terms of the treaty, though practically all the cases tried concerned citizens dwelling within the bounds of the ceded territory.

129

In the Quebec Act of 1774, provision was made for the government of Quebec by civil rule, with the French—not British—laws in force, and this included the Detroit river district. Owing to the distance from Montreal, where all cases were tried, no authority over this district was exercised by anything other than military officers until the proclamation of Lord Dorchester in 1788. The resident Commander of Detroit, exercised the functions both of a military and a civil officer. He married and even baptized those who desired his services, using the forms of the Church of England to conduct the ceremony. In Judicial matters, he exercised the functions of a Judge. The attempt to establish civil rule over a people, though British subjects, dwelling in a land which another nation claimed as theirs, could not reasonably be expected to continue very long.

Circumstances Leading to Renewal of War

A set of circumstances arose in 1793, which brought the matter into the arena of final discussion. Napoleon, the greatest of political adventurers which the world has ever known, and for a time the most successful of military leaders, had commenced war activities in Europe in which Great Britain was compelled to take a part. The resources of the motherland were all, therefore, required for this momentous struggle, a conflict which had to be continued for twenty years before victory finally crowned her efforts.

With the outbreak of this war, the question of the outposts, and the wresting of the whole of Canada from Great Britain, became a live political issue in the United States. The agitation, if not incited, at least received strong support, from the presence in the United States country, of increasing numbers of "United Irishmen," an association, which at that time was giving Great Britain, considerable trouble.*

This was not the ennunciation of a new policy. It was the policy followed out by all European colonization efforts in North America from its earliest beginning. It was force against force with the strongest taking possession of the land. It was not a question of who discovered it, or who occupies it, but who is it that can conquer and hold that which he has conquered. This policy was carried out first in respect to the Indian. When he was removed as a factor opposing occupation, then it became France's turn to protect her discoveries and her occupied territory. In 1760, there was a force out for conquest greater than the French could muster and wrested from them all their possessions in North America. When the British became a divided house, and the Republicans successful in the struggle, then it was between the two branches the same as of old.—He shall have who can take and hold.

But in addition to land lust, there was another impelling force in American politics accounting for their desire to enter into war with Great Britain. This was the influence of the Republican idea upon their ideals and thoughts. Republicanism, at this time, was arresting the attention of the whole of Europe, as well as of North American, and was sweeping over these nations with increasing popularity.

The popularity of this Republican idea, Upper Canada was seeing exemplified every day in that period of its history.

*That the self-created socities which have spread themselves over this country, have been laboring incessantly to sow the seeds of distrust, jealousy, and of course discontent, thereby hoping to effect some revolution in the government, is not unknown to you. That they have been the fomenters of the western disturbances, admits of no doubt in the mind of any who will examine their conduct . . . "
—President Washington to John Jay, Nov. 1st, 1794

Emigrants from England, Scotland, Ireland and Wales were constantly passing through its territory from Niagara to Detroit, yet none of the fertile lands which lay on either side of the trail over which they passed had any influence of attraction for them. Indifferent and unheeding they trudged on to "the land of the free." With such sentiments prevailing among the people of Great Britain, and with an increasing number of British people favouring and favourable to the American Republic, it was an opportune time to extend the boundaries of the United States to include the whole of North America.

Governor Simcoe Would Stand Aside, Neutral

The Indians north of the Ohio, were claiming themselves to be a free and independent nation. Their ownership of the Indian Territory had been recognized by Great Britain, in the terms of both the treaty of Utrecht and the Quebec Act of 1774, a claim which their friends and allies, the fur-traders, were heartily supporting.

Two American armies had been sent out to bring the Miamis into subjection, but both of them suffered terrible defeat. The year 1794 saw another army being mobilized. Governor Simcoe, in view of the warlike preparations going on, deemed it expedient to re-occupy the abandoned British trading post, Fort Miami, situated at the foot of the Miami rapids, thirteen miles inland from Lake Erie. Here one of the most prominent inhabitants and fur-traders of the Detroit river district, Alexander McKee, had his place of business, a trader whom Lord Dorchester had but a short time previously appointed Superintendent of Indian affairs for the district. When friendship with all of the Indians of that neighbourhood had been obtained, the cessation of hostilities between French and British had become a matter of history, there was no need of incurring the expense of maintaining a garrison post there. But now a new set of circumstances had arisen, which seemed to call for its re-occupation. The American Republic had taken up arms against the Indians of the Ohio and the Miami territory and had marched their army forward to give them battle in their own territory.

American Conflict with the Indians

The policy of Governor Simcoe in respect to this conflict was that of strict neutrality.* But this was a position that was going to be very difficult for him to maintain, because not in accord with either the expectations of the Indians, the sympathies of the Canadian fur-traders, or the aggressive

---

*Whatever may have been the influences exerted upon Governor Simcoe afterwards to change his mind in respect to war, it is certain that when he first took up the reins of government, his desire, his hope, and his activities, were all combined in the one aim to secure for the province the blessing of peace. The Americans had received their independence, and much more territory than that to which they were entitled. Canada was started out on its own career. Let both work out their own separate destinies, but let it be done as peaceable neighbours not as warring nations. If the retention of these outposts should lead to war, then let them be given up, even with Detroit, rather than such an eventuality should take place. This was, therefore, his final judgment in respect to the frontier posts.

"I am convinced that my system is just to prevent war, by the appearance of force and by its concentration, I do therefore most earnestly hope that the Companies, a British regiment and a good field train, will be established in my intended capital."

"I mean to state strongly my situation, that I may be fully exculpated to my country in case of any sinister events. The British Regiment and Field Artillery are my request to prevent war . . . I have no personal views, no personal fears, but those of peace, peace, peace . . . if we are forced into war while I govern Upper Canada, it shall not be the wisest sort, preventitive war, but absolutely and entirely defensive, of which the most striking proofs shall be given to mankind; proofs that neither the sophistry of the States nor the credulity of Great Britain can obscure."

attitude of the American Republic.

The Indians of the North West were the allies of Great Britain at this time, and had assisted the Constitutionalists in the American Revolutionary war. They demanded of the Canadians that they show their friendship by taking their side in this war, as they had done for them in the other. With this position, the fur-traders were in sympathy, which placed them at variance with Simcoe's policy of neutrality, and it was impossible to compel them to be neutral either in their thinking or acting.* They wanted the Indians maintained as an independent people, having their own territory, and the right to retain or dispose of it according to their own desire and purpose.

The Americans had already taken a belligerent attitude and were in the Indian country, making gestures of hostility with an armed force under General Wayne. The policy of Governor Simcoe for peace and neutrality could not be maintained if hostilities between the Americans and the Indians should be long continued. He counselled mediation, but the American Republic had determined to rid itself of Indian occupation of the Ohio country, even if it meant the extirpation of the whole race of Indians in that region. Of their settled purpose to carry out that policy, there was no lack of evidence. It was professedly for this end that the army of General Wayne was mobilized and sent into this region.

---

*"I am led by several little circumstances, not easily detailed or explained, to believe that the late administration looked upon war with us as inevitable; and I am of opinion that the instructions of the 6th November were influenced by that idea. I do also believe that Lord Dorchester was instructed to act conformably to that idea, and that Simcoe was governed by it.
—John Jay to President Washington, July 31st, 1794

With the American threat to wrest Canada from Great
Britain, with the re-occupation of Fort Miami by Governor
Simcoe, with General Wayne making gestures of hostility on
the western frontier of the Detroit river district, there was
created a condition of affairs which brought Great Britain
and the United States once more on the verge of another war.
To avert this catastrophe, if possible, President Washington
had previously sent Chief Justice John Jay to Great Britain.
He was deputed to go on an errand of peace, while at the same
time, General Wayne's army continued its activities in the
North West territory.

General Wayne's Activities on the Miami

As Wayne's army moved westward and daily approached
nearer to Fort Miami, the military atmosphere of the Detroit
river district became tense with excitement. The militia of
both Essex and Kent counties were ordered to assemble that
they might be ready for immediate action if the occasion
required it. Thirteen hundred Indians were gathered together
from different tribes and took a position near the Miami
Rapids, in order to intercept and give battle to this westward-
moving force. The engagement took place on the twentieth
day of August, the season of the Indian corn harvest. The
Indians were defeated and scattered.

Immediately following this success, General Wayne turned
his attention towards the Canadian Fort, and demanded the
surrender of the three hundred soldiers manning it. Major
Campbell, the commander, of course refused, and made
preparations for defence. In the exchanges which took place
there was a marked contrast between the two leaders
exhibited. Major Campbell refused, however, to surrender his

dignity, and answered in exchange with the cool deliberation of a brave, cultured and resolute military officer and gentleman.

The demand for surrender was not followed up by an attack to compel compliance. Instead, a council of war was called by General Wayne which concluded that the American army had not with them an equipment of guns adequate for a successful attack. Accordingly, General Wayne halted his march westward, and withdrawing northward, took a position on the Upper Glaize, where he busied himself in the interval in the erection of another fort which he named Fort Defiance. This withdrawal, however, was preceded by the setting on fire and the laying waste of the villages, the cornfields and the haystacks of the Indians inhabiting the district.

The Wanton Destruction of Indian Agent McKee's Property

In this wanton waste, Alexander McKee, whose place of business was established in the immediate neighbourhood of Fort Miami, shared a similar fate. What justification could be given for the destruction of this civilian's property? Born on North American soil, a native of Pennsylvania, he had committed the crime of being a Constitutionalist American, and supported that party in the ranks of their army. This war was now over, but not the antipathies engendered by it. "He's our old-time enemy, and we must punish him now that we have the chance." This explains the deed, but does not justify it. If further motive were required, it would be found in the fact that he had become, along with others of his countrymen, a Canadian, desirous of building up in the domain of Canada, a country according to his predilections, independent in its government, but British in its national connections, exemplifying the ideals for which he fought in the Constitutionalist army of America. But in the destruction of his property, he escaped better than he anticipated. It is on record that the day previous to the battle, he made his will, anticipating that his person, as well as his property, would be in imminent danger at the hands of the approaching army.

Canadian Militiamen Assist the Indians

In this battle, there was another incident occurred, which in itself was a matter of no great moment, but which connected with the international circumstances of the times, was deemed an event of maximum importance, the presence of a small body of Canadians with the Indians in the battle. The policy of Governor Simcoe, strict neutrality in respect to the conflict between the Americans and the Indians, was, as we have observed, at variance with the sympathies of British Canadians, especially those in the fur-trade. They reasoned that the Canadians had a right to support the Indians. They are our allies and a part of the defensive forces of the country. In the event of a war with the American Republic, and everyone believed that such a war was imminent, the Indian would be asked and expected to aid the Canadian cause. Only with such aid could the independence and the integrity of the Canadian domain be maintained. Shall we permit, they asked, the destruction and extirpation of the Indian, a part of our own defensive army, without now, lifting so much as a finger to aid him, when he is in need?

How strongly the British Canadians felt concerning this matter can be gauged from the fact, that, notwithstanding their knowledge that it was against the policy of their government, certain of these took their lives in their hands, and stood side by side with the Indians fighting at the Miami Rapids in defence of Indian national rights and Indian territory. Among these was an officer of the Canadian militia, Captain Caldwell, a circumstance which gave rise to the complaint that the presence of the Canadians in that battle was due to Government sympathy and authority.

Amongst our own Canadian population, there was not a unanimous opinion as to the legitimacy of Captain Caldwell and his associates taking part in that battle. Those who were intimidated by Wayne's display of American strength, or who were indifferent as to the need of protecting this part of the Canadian domain, believing it, like Oswald, as a territory of no value, these looked upon their action as an exhibition of great imprudence, perhaps even an international crime. But those who believed that the Canadian domain should remain Canadian, and be saved from partition, these looked upon the aid given by them to the Indian as an act of heroism, a heroism in which every eligible Canadian who could have been mustered for the occasion, ought to have taken a part.

Governor Simcoe's Military Prescience

There was, perhaps, room for a difference of opinion, as to whether or not Captain Caldwell, being an officer of the Canadian militia, should have acted the part that he did, without the consent and authority of his superior officers. On the other hand, that any informed British-man, dwelling in either Canada or the Old Country, should attach any blame to Governor Simcoe for his re-habilament of Fort Miami is indeed hard to understand. Yet his act has been designated as one of hostility towards the American Republic, displaying a belligerent spirit, when he ought to have been sowing the seeds of good-will and cultivating friendship with them.

The unwisdom or wisdom of Governor Simcoe's act in re-occupying Fort Miami depends altogether upon the reason accounting for the presence of the American army in the western country. When General Wayne moved his army into a territory that was indisputably a part of the Canadian domain in the French regime, and the cession of which to the the United States was still a matter under discussion, had the documentary evidence on his person that this territory had been ceded over to the American Republic in the same way that Major Rogers had from Governor Vaudreuil, when he came to receive the surrender of Fort Detroit from the commandant, Colonel Belestre? No official document had come from the British Government authorizing evacuation. Until such official notification arrived, Governor Simcoe must treat that territory as he would any other part of the province over which he was appointed Governor, and provide for its inhabitants that measure of protection for their persons and their property which was within his power, and this until such time as he received orders, not from General Wayne but from the British Government, to evacuate the posts and the territory to be ceded over to the United States.

For eleven years they had refused to issue that order, and had forwarded to the American Government the reasons for their refusal. They held back, they claimed, in order to insure the payment of the just claims which they had in respect to moneys and property against the United States. If these claims were fictitious, then neither Chief Justice John Jay nor President George Washington would have acknowledged them nor deemed themselves under any obligation to pay them. In a letter sent by Mr. Jay to the President, he puts himself clearly on the side opposed to the confiscation of British claims and the cancellation of United States debts to Great Britain.

"I learn that Virginia is escheating British property, and I hear of other occurrences which I regret; but they shall not abate my perseverance in endeavouring to prosecute peace, and bring the negotations to such a conclusion as will either insure peace with this country, or produce union among ourselves in prosecuting war against it."*

President Washington Smooths the Way for Peace

To President Washington, much praise has been given for his attitude in relation to these disputed claims. Mr. Jay mentions Virginia as one of the States which refused to consider these claims and rights as sacred. The same was true of other States also. In Massachusetts, John Powell, the father of the pioneer Judge of Upper Canada, although a native of the State, having moved his famliy to England when a clash between the two contending parties seemed imminent, was forbidden to return and his properties were declared forfeited.

---

*Your letter "of the 5th of August dawns more favourably upon the success of your mission than any that preceded it, and for the honour, dignity, and interest of this country; for your own reputation and glory; and for the peculiar satisfaction I should desire from it, as well on private as on public considerations, no man more ardently wishes you complete success than I do. But, as you have observed in some of your letters, that it is hardly possible in the early stages of negotiation to forsee all the results, so much depending upon fortuitous circumstances, and incidents which are not within our control,—so to deserve success, by employing the means with which we are possessed to the best advantage, and trusting the event to the all-wise Disposer, is all that an enlightened public, and the virtuous and well-disposed part of the community can reasonably expect; nor in this, will they, I am sure, be disappointed. Against the malignancy of the discontented, the turbulent, and the vicious, no abilities, no exertions, nor the most unshaken integrity are any safeguard."
—President Washington to John Jay, November 1st, 1794

There were thousands of other similar cases. All of those rights and claims antedating the war, Great Britain stipulated should be recognized and adjusted. This being agreed upon, it only remained that the ceded territory should be handed over to the United States.

Although these claims ran up to millions of dollars in value,—six hundred thousand pounds was paid by the United States in 1802—yet they were as nothing in worth compared with the value of the ceded territory. The friendship and trade of Great Britain was of greater value to the United States than were the property issues at stake. Similarly the friendship and trade of the United States was worth far more to Great Britain than the fur-trade and its protecting outposts in the ceded territory. From an economic viewpoint, if there were no other, both nations would be benefitted by the removal from the arena of controversy, policies which were doing neither any good, but provoking both to war.

Chief Justice John Jay is the one man above all others to whom credit is due for bringing about a peaceable settlement between them. It is the first instance of a settlement on the basis of negotiation and compromise, a method which the good graces and common sense of both nations have consistently followed on every, save one, occasion since in the settlement of their international problems. In sending Chief Justice John Jay as the apostle of good-will to Great Britain, in the judgment of President Washington, he was sending the best possible man available, one to whom, in the formation of his government, he had offered in it any office he cared to choose. But in understanding to carry through a measure which would recognize justice in the terms of its settlement, Mr. Jay was assaying a task that would in its unpopularity with a section

of his own people overshadow the influence which his reputed
worth might be expected to have on the generosity of their
thinking.

Chief Justice John Jay, Plenipotentiary

When he arrived in London, he met there a friendly
government. Apart altogether from the fact that their hands
were tied with one war, and therefore with no inclination to
enter into a second, the Grenville ministry represented that
part of the British people who were friendly disposed towards
the United States. These did not consider the question at issue
of sufficient importance to justify war. Besides they were
now directing their aspirations seaward. Their faces looked
forward to the time when their fleet of mercantile vessels
would be amply protected by an equal number of war vessels.
To the accomplishment of this end they were bending their
energies, purposing to make of the British Isles a nation of
sailors, not of farmers. In the face of these mercantile and
naval aspirations, there was not found a sufficient interest
in the Indian country north of the Ohio to support any
government that might be in power in Great Britain, to
prosecute war against the American Republic in order to
retain it as Indian possession.

In conducting the negotiations, Mr. Jay was given wide
latitude, his instructions from the Government being more in
the form of recommendations, it being expected that he would
use his own best judgment in effecting a settlement of the
disputed issues. Lord Grenville was appointed by the King
as the representative of Great Britain. To these two were
given the power to effect a treaty respecting three matters,—

1. A settlement of all existing disputes in relation to

the treaty of peace, 1783.

2. A treaty of commercial rights and privileges to be granted to the United States for trade with Great Britain and its colonies.

3. Compensation to American merchants for losses sustained through the infringement of the rights of neutral nations in connection with the Napoleonic war.

It is only with the first of these three that the interests of the Detroit river district were directly connected. Concerning the outposts, it was agreed that they should be evacuated on or before June 1st, 1796. The settlers and traders within the jurisdiction of the posts were to be permitted to remain and to enjoy their property without becoming citizens of the United States, unless they should think proper to do so. Between the United States and the Canadians, it was agreed that the inhabitants might freely pass, by land or inland navigation, into the territory of the two parties, and carry on trade with each other; and that the duties on goods thus imported should be the same as those paid by the citizens of the country into which they were imported. Lands held by the subjects of the two parties in the territories of the other, were confirmed to them and their heirs.

Reception Accorded His Successful Undertaking

In effecting a settlement, compromises had of necessity to be made on both sides. In this way, this first official endeavour to establish mutual good-will between the two countries, happily resulted in successful accomplishment. The treaty was signed on the 19th of November, 1794; ratified by the American Senate the following Spring, (one clause in reference to trade excepted) ; signed by President Washington, August 15th, 1795; and declared on March the 3rd, the following year, to be the supreme law of the land.

By a section of the United States people, it was bitterly opposed. Chief Justice Jay was attacked as a renegade traitor to his country and burned in effigy; his friend, Alexander Hamilton, was stoned at a public meeting for speaking in favour of it; and it was saved from rejection only by the firmness and courage of President Washington. This, Chief Justice Jay himself anticipated before he undertook the task, knowing full well that there was a section of extremists in the country who wanted war with Great Britain and would not be satisfied with anything else.* Yet, notwithstanding its unpopularity, this successful effort in averting war, and introducing to a war-loving world the method of arbitration as a proper mode of settling international disputes, is recorded in the chronicles of American history, as the most well-known to-day of his life's achievements.

"I cannot conclude this letter without repeating to you," wrote Lord Grenville, his colleague, when commending Chief

---

*I am certain that intelligence (which made some impression) was conveyed to the ministry, that our army, if successful against the Indians, had orders to attack and take the posts. There is also room to believe, that the indiscreet reception given to the late French minister— the unnecessary rejoicings about French success, and a variety of similar circumstances, did impress the government with strong apprehensions of an unavoidable war with us, and did induce them to entertain a disposition hostile to us.

I have given Lord Grenville positive assurances, that no attack pending the negotiations will be made on the posts held by them at the conclusion of the war; but I also told him that I thought it highly probable that every new advanced post, and particularly the one said to be taken by Mr. Simcoe on the Miami, would be attacked. I must do him the justice to say, that hitherto I have found him fair and candid, and apparently free from asperity or irritation."
—Letter John Jay to President Washington, July 21st, 1794.

ST. JOHN'S CHURCH, SANDWICH

This congregation of the Church of England, is the pioneer Protestant congregation of the Detroit river district. It was established by Richard Pollard, the sheriff of the Western District, ordained a clergyman, March 21st, 1802. The cemetery attached to it, is the burying place of many of the prominent pioneers of the first period in the history of the district.

A small log building was put up in 1796, where Mr. Pollard, then a layman held services. He became first rector when ordained in 1802. This church was destroyed by fire September 13th, 1813.

Justice Jay's part in the negotiations, "the very great satisfaction I have derived from the open and candid manner in which you have conducted on your part the whole of the difficult negotiation, which we have now brought to so successful an issue, and from the disposition you have uniformly manifested, to promote the objects of justice, conciliation and lasting friendship between the two countries."

In Canada, although its reception elicited no enthusiasm, yet there was no strong sentiment against it. The calmer minds of both countries accorded their pleasure that an eleven-year old controversy had come to a peacable and cordial end, reflecting credit not only to both of the men negotiating the terms, but to the common sense of the two nations as well.

One hundred and thirty five years have elapsed since the movements of these men with their rival policies took place on the checker board of fate. By this treaty, the destined area of Upper Canada was determined in respect at least to its south-western boundaries. We are interested, however, at this period in the history of the province, not in what was given away, but in what was left. Even with this partition taking place, Upper Canada has still remaining a territory in area equal to that of the three great nations of Europe—the British Isles, France and Italy—combined. There should therefore be no complaints nor regrets because of what took place in the distant past when we realize what a heritage of possibilities still remaineth to us.

For the evacuation of the outposts, two years of respite was given, so that the treaty though made in 1794 did not go into effect until 1796, at which period the Canadian Detroit river district became confined to that section of the territory lying on its eastern shore.

In accord with the terms of the treaty the evacuation should have taken place on June 1st, but owing to unexpected circumstances it did not actually take place for a couple of months later. In the meantime, preparations for the evacuation commenced in 1794. The expropriation of land for a town-site at the mouth of the Detroit River, where was to be established a military post and naval station, was the first undertaking set in motion, in order to affect removal from Detroit within the time-limit allowed. The second was the establishment of a jail and court-house at Sandwich and such other preparations as were necessary in order to make it the capital town of the district. The first to change their residences to the Canadian shore were the traders, most of whom settled at Amherstburg or Fort Malden, though many moved north to Fort St. Joseph. One, Angus McIntosh, established a trading-post about three miles east of Sandwich, where is now the boundary line between the town of Walkerville and the city of Windsor, which he named Moy. This building remained until very recent times an historic landmark, reminiscent of those pioneer days in the history of the Detroit River region. A goodly number of the British, and practically all of the French-Canadians, continued their residence in the State of Michigan. The terms of the Treaty permitted them to remain and enjoy possession of their property and freedom for trade without being under the necessity of taking the oath of allegiance to the United States. The Jay Treaty was passed with a view to establish friendly relationships among all dewelling on either side of the Detroit River. Unfortunately, this anticipation was unrealized, the War of 1812 shattered this fabric of hope and created embittered feelings even between members of the same household, which it took years after to wholly eliminate.

CHAPTER XII.

The Ceded Outposts

A Contemporary Estimate of
Their Value

I.

THE American prints, until the late treaty of amity was ratified, teemed with the most gross abuse of the British government, for retaining possession of Niagara Fort, and the other military posts on the lakes, after the independence of the States had been acknowledged, and peace concluded. It was never taken into consideration, that if the British government had thought proper to have withdrawn its troops from the posts at once, immediately after the definitive treaty was signed, the works would in all probability have been destroyed by the Indians, within whose territories they were situated, long before the people of the States could have taken possession of them; for no part of their army was within hundreds of miles of the posts, and the country through which they must have passed in getting to them was a mere wilderness; but if the army had gained the posts, the states were in no condition, immediately after the war, to have kept in them such large bodies of the military as would have been absolutely necessary for their defence whilst at enmity with the Indians, and it is by no means improbable, but that the posts might have been soon abandoned. The retention of them, therefore, to the present day, was, in fact, a circumstance

147

highly beneficial to the interests of the States, notwithstanding that such an outcry was raised against the British on
that account, inasmuch as the Americans now find themselves
possessed of extensive fortifications on the frontiers, in perfect
repair, without having been at the expense of building them,
or maintaining troops in them for the space of ten years,
during which period no equivalent advantages could have been
derived from their possession.   It is not to be supposed,
however, that the British government meant to confer a
favour on her late colonies by retaining the posts; it was well
known that the people of the new states would be eager sooner
or later, to get possession of forts situated within their
boundary line, and occupied by strangers; and as there were
particular parts of the definitive treaty which some of the
states did not seem very ready to comply with, the posts were
detained as a security for its due ratification on the part of
the States.   In the late treaty of amity and commerce, these
differences were finally accommodated to the satisfaction of
Great Britain, and the posts were consequently delivered up.
On the surrender of them very handsome compliments were
paid, in the public papers throughout the States, to the British
officers, for the polite and friendly manner in which they
gave them up.   The gardens of the officers were all left in full
bearing, and high preservation; and all the little conveniences
were spared, which could contribute to the comforts of the
federal troops.

The generality of the people of the States were big with
the idea, that the possession of these places would be attended
with the most important and immediate advantages; and in
particular they were fully persuaded that they would thereby
at once become masters of the trade of the lakes, and of three-

fourths at least of the fur-trade, which, they said, had hitherto been so unjustly monopolized by the British merchants, to their great prejudice. They have now got possession of them, and perceive the futility of all these notions.

The posts surrendered are four in number; namely, Fort Oswego, at the mouth of Oswego River, which falls into Lake Ontario, on the south side; Fort Niagara, at the mouth of Niagara River; Fort Detroit, on the western bank of Detroit River: and Fort Michilimackinac, at the straits of the same name, between Lake Michigan and Lake Huron. From Oswego the first of these, we derived no benefit, whatever. The neighbouring country, for miles round, was a mere forest; it was inhabited by few Indians, and these few carried their furs to Cadarague or Kingston, where they got a better price for them than at Oswego, as there were many traders there, and of course some competition amongst them; at the same time, the river, at the mouth of which this fort stands, was always open to the people of the States, and along it a small trade was carried on by them between New York and Lake Ontario, which was in no wise even interrupted by the troops at the fort. By the surrender of this place, therefore, they have gained nothing but what they enjoyed before, and the British government is saved the expense of keeping up a useless garrison of fifty men.

The quantity of furs collected at Niagara is considerable, and the neighbourhood being populous, it is a place of no small trade; but the town, in which this trade is carried on, being on the British side of the line, the few merchants that lived within the limits of the fort immediately crossed over to the other side, as soon as it was rumoured that the fort was to be given up. By the possession of a solitary fort, therefore, the people of the States have not gained the smallest portion of this part of the lake trade; nor is it probable that any of them will find it their interest to settle as merchants near the fort; for the British merchants, on the opposite side, as has already been shown, can afford to sell their goods, brought up the St. Lawrence, on much lower terms than what goods brought from New York can be sold at; and as for the collecting of furs, it is not to be imagined that the Indians, who bear such a rooted hatred to the people of the States, who are attached to the British, and who are not a people ready to forsake their old friends, will carry their furs over to their enemies, and give up their connections with the men who can afford to pay them so much better than the traders on the opposite side of the water.

Detroit, of all the places which have been given up, is the most important; for it is a town, containing at least twelve hundred inhabitants. Since its surrender, however, a new town has been laid out on the opposite bank of the river, eighteen miles lower down, and hither many of the traders have removed. The majority of them stay at Detroit; but few or none have become citizens of the States in consequence, nor is it likely that they will, at least for some time. In the late treaty, a particular provision for them was made; they were to be allowed to remain there for one year, without being called on to declare their sentiments,* and if at the end of that period

---

*This part of the late treaty has by no means been strictly observed on the part of the States. The officers of the federal army, without asking permission, and contrary to the desire of the remaining British inhabitants, appropriated to their own use several of the houses and stores of those who had removed to the new town, and declared their determination of not becoming citizens of the States; and many of the inhabitants had been called on to serve in the militia, and to perform duties, from which as British Subjects, they were exempted by the articles in the treaty in their favour. When we were at Detroit, the British inhabitants met together, and drew up a memorial on the subject, reciting their grievances, which was committed to our care, and accordingly presented to the British minister at Philadelphia.

they chose to remain British subjects, they were not to be molested in any manner, but suffered to carry on their trade as formerly in the fullest extent; the portion of the fur trade, which we shall lose by the surrender of this place, will therefore be very inconsiderable.

The fourth post, Michilimackinac, is a small stockaded fort, situated on an island. The agents of the North-west Company of merchants at Montreal, and a few independent traders, resided within the limits of the fort, and bartered goods there for furs brought in by different tribes of Indians, who are the sole inhabitants of the neighbouring country. On evacuating this place, another post was immediately established, at no great distance, on the Island of St. Joseph, in the Straits of St. Mary, between Lakes Superior and Huron, and a small garrison left there, which had since been augmented to upwards of fifty men. Several traders, citizens of the States, have established themselves at Michilimack-inac; but as the British traders have fixed their new post so close to the old one, it is nearly certain that the Indians will continue to trade with their old friends in preference, for the reasons before mentioned.

From this statement it appears evident, that the people of the States can only acquire by their new possession a small part of one branch of the fur trade, namely, of that which is carried on one of the nearer lakes. The furs brought down from the distant regions in the north-west to the grand portage, and from thence in canoes to Montreal along the Ottawa River, are what constitute by far the principal part, both as to quantity and value, of those exported from Montreal; to talk, therefore, of their acquiring possession of three-fourths of the fur trade by the surrender of the posts on the lakes is absurd in the extreme; neither is it likely that they will acquire any considerable share of the lake trade in general, which, as I have already pointed out, can be carried on by the British merchants from Montreal to Quebec, by means of the St. Lawrence, with such superior advantage.

It is worthy of remark, that as military posts, all those lately established by the British are far superior, in point of of situation, to those delivered up. The ground on which the new block house is building, on the British side of Niagara River, is nine feet higher than the top of the stone house in the American fort, and it commands every part of the fort. The chief strength of the old fort is on the land side; towards the water, the works are very weak, and the whole might be battered down by a single twelve pounder judiciously planted on the British side of the river. At present it is not proposed to erect any other works on the British side of the river than the block house; but should a fort be constructed hereafter it will be placed on Mississaugis Point, a still more advantageous situation than that on which the block house stands, as it completely commands the entrance into the river.

The new post on Detroit River commands the channel much more effectually than the old fort in the town of Detroit; vessels cannot go up or down the river without passing within a very few yards of it. It is remarkable, indeed, that the French, when they first penetrated into this part of the country, fixed upon the spot chosen for this new fort, in preference to that where Detroit stands, and they had absolutely begun their fort and town, when the whole party was unhappily cut off by the Indians.

The island of St. Joseph, in the third place, is a more

eligible situation for a British military post than Michilimackinac, inasmuch as it commands the entrance of Lake Superior, whereas Michilimackinac only commands the entrance into Lake Michigan, which is wholly within the territory of the United States.

It is sincerely to be hoped, however, that Great Britain and the United States may continue friends, and that we never may have occasion to view those posts on the frontiers in any other light than as convenient places for carrying on commerce.

II.

The Ceded Town of Detroit

A Description of the Town which the British Civic Officials and Garrison Soldiers Evacuated in 1796

WE remained for a short time in Malden, and then set off for Detroit in a neat little pleasure boat, which one of the traders obligingly lent us. The river between the two places varies in breadth from two miles to half a mile. The banks are mostly very low, and in some places large marshes extend along the shores, and far up into the country. The shores are adorned with rich timber of various kinds, and bordering upon the marshes, where the trees have full scope to extend their branches, the woodland is very fine. Amidst the marshes, the river takes some very considerable bends, and it is diversified at the same time with several large islands, which occasion a great diversity of prospect.

Beyond Malden no houses are to be seen on either side of the river, except indeed the few miserable little huts in the Indian villages, until you come within four miles or thereabouts of Detroit. Here the settlements are very numerous on both sides, but particularly on that belonging to the British. The country abounds with peach, apple, and cherry orchards, the richest I ever beheld; in many of them the trees, loaded with large apples of various dyes, appeared bent down into the very water. They have many different sorts of excellent apples in this part of the country, but there is one far superior to all the rest, and which is held in great estimation, called the pomme caille. I do not recollect to have seen it in any other part of the world, though doubtless it is not peculiar to this neighbourhood. It is of an extraordinary large size, and deep red colour; not confined merely to the skin, but extending to the very core of the apple: if the skin be taken off delicately, the fruit appears nearly as red as when entire. We could not resist the temptation of stopping at the first of these orchards we came to, and for a few pence we were allowed to lade our boat with as much fruit as we could well carry away. The peaches were nearly out of season now, but from the few I tasted, I should suppose that they were of a good kind, far superior in flavour, size, and juiciness to those commonly met with in the orchards of the middle states.

The houses in this part of the country are all built in a similar style to those in Lower Canada; the lands are laid out and cultivated also similarly to those in the lower province; the manners and persons of the inhabitants are the same; French is the predominant language, and the traveller may fancy for a moment, if he pleases, that he has been wafted by enchantment back again to the neighbourhood of Montreal or Three Rivers. All the principal posts throughout the western country, along the lakes, the Ohio, the Illinois, &c. were established by the French; but except at Detroit and in the neighborhood, and in the Illinois country, the French settlers have become so blended with the greater number who spoke English, that their language has every where died away.

Detroit contains about three hundred houses, and is the largest town in the western country. It stands contiguous to the river, on the top of the banks, which are here about twenty feet high. At the bottom of them there are very extensive wharves for the accommodation of the shipping, built of wood, similar to those in the Atlantic sea-ports. The town consists of several streets that run parallel to the river, which are intersected by others at right angles. They are all very narrow, and not being paved, dirty in the extreme whenever it happens to rain; for the accommodation of passengers, however, there are footways very close to each other. The town is surrounded by a strong stockade, through which there are four gates; two of them open to the wharfs, and the two others to the north and south side of the town respectively. The gates are defended by strong block houses, and on the west side of the town is a small fort in form of a square, with bastions at the angles. At each of the corners of this fort is planted a small ordnance at present in the place. The British kept a considerable train of artillery here, but the length of time against a regular force; the fortifications, indeed, were constructed chiefly as a defence against the Indians.

About two thirds of the inhabitants of Detroit are of French extraction, and the greater part of the inhabitants of the settlements on the river, both above and below the town, are of the same description. The former are mostly engaged in trade, and they all appear to be much on an equality. Detroit is a place of very considerable trade; there are no less than twelve trading vessels belonging to it, brigs, sloops and schooners, of from fifty to one hundred tons burthen each. The land navigation in this quarter is indeed very extensive, Lake Erie, three hundred miles in length, being open to vessels belonging to the port, on the one side; and Lakes Michigan and Huron, the first upwards of two hundred miles in length, and sixty in breadth, and the second, no less than one thousand miles in circumference, on the opposite side; not to speak of Lake St. Clair and Detroit River, which connect these former lakes together, or of the many large rivers which fall into them. The stores and shops in the town are well furnished, and you may buy fine cloth, linen, &c. and every article of wearing apparel, as good in their kind, and nearly as reasonable terms, as you can purchase them at New York or Philadelphia.

The inhabitants are well supplied with provisions of every description; the fish in particular, caught in the river and neighbouring lakes, are of a very superior quality. The fish held in most estimation is a sort of large trout, called the Michilimackinac white fish, from its being caught mostly in the straits of that name. The inhabitants of Detroit and the neighbouring country, however, though they have provisions in plenty, are frequently much distressed for one very necessary concomitant, namely, salt. Until within a short time past they had no salt but what was brought from Europe; but salt springs have been discovered in various parts of the country, from which they are now beginning to manufacture that article for themselves. The best and most profitable of the springs are retained in the hands of government, and the profits arising from the sale of the salt are to be paid into the treasury of the province. Throughout the western country they procure their salt from springs, some of which throw up sufficient water to yield several hundred bushels in the course of one week.

There is a large Roman Catholic church in the town of Detroit, and another on the opposite side, called the Huron church, from its having been devoted to the use of the Huron Indians. The streets of Detroit are generally crowded with Indians of one tribe or other. At night all the Indians, except such as get admittance into private houses, and remain there quietly, are turned out of the town, and the gates shut upon them.

The country around Detroit is very much cleared, and so likewise is that on the British side of the river for a considerable way above the town. The settlements extend nearly as far as Lake Huron; but beyond the River La Tranche, which falls into Lake St. Clair, they are scattered very thinly along the shores. The banks of the River La Tranche, or Thames, as it is now called, are increasing very fast in population, as I before mentioned, owing to the great emigration thither of people from the neighbourhood of Niagara, and of Detroit also since it has been evacuated by the British. We made an excursion, one morning, in our little boat as far as Lake St. Clair, but met with nothing, either amongst the inhabitants, or in the face of the country, particularly deserving of mention. The country round Detroit is uncommonly flat, and in none of the rivers is there a fall sufficient to turn even a grist mill. The current of Detroit river itself is stronger than that of any others, and a floating mill was once invented by a Frenchman, which was chained in the middle of that river, where it was thought the stream would be sufficiently swift to turn the water wheel: the building of it was attended with considerable expense to the inhabitants, but after it was finished it by no means answered their expectations. They grind their corn at present by wind mills, which I do not remember to have seen in any other part of North America.

The soil of the country bordering upon Detroit River is rich though light, and it produces good crops both of Indian corn and wheat. The climate is much more healthy than that of the country in the neighbourhood of Niagara River; intermittent fevers, however, are by no means uncommon disorders. The summers are intensely hot, Fahrenheit's thermometer often rising above 100; yet a winter seldom passes over but what snow remains on the ground for two or three months.

———————

The above estimate of the value of the ceded outposts, together with the description account of Detroit, is a Reprint from a valuable book of travels, written by Isaac Weld, Junior, who visited the Detroit region and wrote an account of his observations during the years VGTD to VGTF. The book is not now in circulation, and the number of available volumes, rare. There is a special interest attached to this narrative account of his visitations, seeing that they are the observations of an eye-witness concerning circumstances and places of which we have not too much contemporary history. His judgments, the circumstances of subsequent years have fully substantiated. ..The fur-trade was not alienated from Canada by the cession of the out-posts. Many of the Indians, hitherto domiciled in the United States, found new hunting grounds in Canada, and have during the years since maintained intact the fur-trade, as a valuable enterprise of our country.—Editor.

CHAPTER XIII.

Ontario's First Farming Community

The Canadian Branch of the Pioneer French Settlement
Of the Detroit River District, Assumption
And Petite Cote

ASSUMPTION was the name given to the first settlement of farmers established in the province of
Ontario. This settlement was situated on the left
bank of the Detroit river, stretching eight miles from the
Canard river northward. The name was first applied, not to
the settlement, which comprised two divisions—Assumption
and Petite Cote—but to the parish, the religious organization
established by the Roman Catholic church among the French
Canadians dwelling on the opposite side of the river from the
town of Detroit. As the oldest settlement, it is the connecting
link between the Ontario of to-day with pre-historic Canada.

The Mission of Recollet Transferred to the Detroit

In the march of circumstance, this religious organization
was a continuing part of the missionary work opened out by
the Recollet priest, Father Le Caron, in 1615, among the
Huron Indians then dwelling in many villages in that section
of the country lying between the Georgian Bay and Simcoe
Lake. On its civic side, it was a part of the colony established
by Cadillac in the Detroit river district in 1701. The work

which Father Le Caron began amongst those Hurons in 1615, was being continued by Fathers Lalemant, Brebeuf and Daniel, when the fierce and warring Iroquois nations made their well-known and successive attacks upon them in 1648 and 1649, and well nigh annihilated the whole number of this tribe among the Indian nations.   Father Daniel suffered martyrdom in the first of these attacks and Fathers Brebeuf and Lalemant in the second.

A remnant of the tribe saved, fled first to the Lake Superior district, where they lived for a time under the shelter of their kinsmen, the Ottawas, but later, they migrated to the Detroit district, and pitched their wigwams permanently in that neighbourhood. They were known under the name of Wyandots, though the name Huron was carefully retained by the French missionaries who later laboured amongst them.

Reverend Father Richardie, Pioneer Misionary

The foundation work, out of which grew the present-day ecclesiastical organization, of Assumption, on the Detroit river, was commenced by Reverend Father Richardie in 1728, the pioneer missionary to whom belongs the honour of establishing the parish of Ste. Anne's in Detroit. He began his activities among the Hurons, or Wyandots, when they were located at the Bois Blanc island at the mouth of the Detroit river. Here he established a mission farm and erected a mission house but no church apparently. After a period of about sixteen years among them there, the Hurons moved from the island and took up their abode on the mainland at a reservation apportioned for them in Sandwich township, and known as Montreal Point.

THE CHURCH OF ASSUMPTION

The Catholic Church established amongst religious institutions the pioneer structure of the Canadian Detroit river district, of which this edifice represents the third stage in the line of its continued progress.

By this time the inertia of old age was beginning to have a marked effect upon the aspirations and efforts of Father Richardie. He wrote to Quebec for an assistant, and one was sent in the person of Father Peter Potier, a native of Belgium. Both of these men took up the work together among the Hurons at Montreal Point, working this in conjunction with Ste Anne's parish in Detroit. Father Richardie retired in1753, and took up his residence at Hotel Dieu, Quebec, where he died five years later at the age of seventy-seven, "full of years and honour."

Parish of Assumption Established

Left to himself, Father Potier's first undertaking was the erection of a church and mission house at Montreal Point. A section of land comprising three hundred and fifty acres, was donated by the Huron Indians from their reservation to this Mission. Upon this he built the first church erected on the Canadian shore. By the time that this forward step was taken in the work among the Indians, a community of Frenchmen had begun to establish themselves on the left bank of the river. Assisted by the home government, farmers from Brittany and Gascony, France, and disbanded soldiers of the French army, were brought out and given land on both sides of the river. Farm implements, grain and other advances were given by the government, who continued this aid until they were able to take care of themselves. The first of these immigrants arrived in 1748. By 1752, twenty families had taken up land and were settled on the Canadian shore. The looking after the spiritual needs of these became an added part of the work of Ste Anne's parish, with which organization they were ecclesiastically connected, a connection which was continued for a period of eight years. During all this period, the French were still masters of the land.

With the change in the national destiny of the district in 1760, there followed also a change in the ecclesiastical arrangements affecting the community on the left bank of the river. By this time there were fifty families located on the south shore, making a population of about three hundred and fifty. This section of the colony was separated from Ste Anne's parish, and connected with the Huron Mission, forming a new organization to which was given the name, the parish of Assumption. This parish became a self-sustaining mission from the first. Hitherto, the French government assisted the Mission, a policy which ceased under British administration. The maintenance of the parish was now made to depend on the aid supplied by the parishioners alone. Father Potier was released from connection with Ste Anne's parish, and given jurisdiction over the parish of Assumption. As this settlement of French Canadians comprised the first farming community of Upper Canada, now Ontario, so this parish was the first ecclesiastical organization also. Their first records are dated July 16, 1761, and as they continue in unbroken sequence ever since, they comprise in the history of the country, "the oldest and most complete file of church records in Ontario."

In 1799, the Huron Indians became separated from the parish, by their removal to another reserve apportioned for them on the Canard river. This left the parish a purely French Canadian one, but to which, later, other nationalities of the Roman Catholic faith were added as the country became more settled, and an urban centre established at Sandwich on the Canadian side of the river.

This little French Canadian colony occupies, then, a unique place in the early history of this province. Their farms

were surveyed two arpents wide facing the river and stretching backward at right angles to the river forty arpents
in length. Their houses were well built log structures, neatly
dove-tailed at the corners, sometimes hewn and sometimes left
in the rough, but in most cases, whitewashed both inside and
out. These were erected a story and a half in height, the roof
covered with split, hand-made shingles, with dormer windows
on the roof facing the river. At the front of the house was the
garden where, in addition to many varieties of flowers, the
vegetables for domestic use were grown. The orchards were
set behind, and included such fruits as peaches, pears, plums
and apples. Roses, lilacs, and hollyhocks were grown extensively on the farm lots, so that at certain seasons of the
year the River Detroit was banked all the way from Lake St.
Clair to the Canard by one continuous line of gardens. Early
travellers are generous in their praises of the beauty in
appearance of these homes. The houses being set within 385
feet, (two arpents), of one another and sometimes two houses
on the one holding, the river bank displayed the appearance
of one continuous village during its whole length. The clearances were very small, or at least not very far distant backward, and behind, a forest which the early missionaries described as 'the finest in the world'.

Nature and art were therefore combined to give an attractive appearance to this first parish, first colony, first
settlement of the province and which, provided as it was, with
so fertile a soil and so mild a climate, with industrious progress,
might have become the first in prosperity also.

Another feature giving character and appearance to these
farm lots was the presence of a cross erected on the front of
each home at the request and under the instruction of their
missionaries.

In the establishment of the Detroit colony, the feudal system of land ownership practised on Old France was followed. Roughly speaking the social fabric there was divided into four grades: first, the king; second, the nobles, the dukes, counts, viscounts, barons, marquises, etc.; third, the seigniors; and fourth, the vassals or fiefs. The authority and influence of these corresponded to the grade in the social fabric which they represented. The king was first in authority, and all of the others followed in a descending scale. The ownership of the land was primarily vested in the king, and he not only owned the land but in a sense the people also.

The land was divided up first among the nobles under the king, unto whom they pledged loyalty and fidelity, as well as payment of taxes, military service and such other support of their sovereign as might be required of them. These again parcelled out the land to the next grade below them, the seigniors, who in turn pledged to the nobles the same fealty as they to their king. Before the fourth-grade man obtained the privilege of tilling the soil, he was compelled to enter into an agreement to fulfill all the requirements demanded of him by the seignior. Thus the vassal paid his dues to the seignior, the seignior to the noble and the noble to the king. In short, the tillers of the soil had to support themselves and the three grades above them, and in the event of failing to do this, the punishment was the loss of their land, which meant of necessity, the loss of their living.

Only two of these four grades took up their residence in Canada—the seigniors and the vassals, with the ownership of the land vested in the king. The rank of Cadillac was that of a seignior, though he was charged with aspirations to be appointed a noble with the title of Marquis. But if he ever had such an ambition, he not only failed to realize it, but even after a period of ten short years, his position of Commandant was taken from him, his property rights were annulled, and his removal to Louisiana ordered by the king. This the reward which the world too often accords the men who serve it best.

As to Cadillac there came no reward of personal wealth as the result of his worth while endeavour, so to the hubsandman. Increase of clearance while it would provide for him larger harvests, would also require of him the payment of increased dues to be paid the seignior. There was nothing in the system to inspire a man to industry and thrift, or induce him to seek higher standards of prosperity and comfort.

But if a too easy contentment with the primitive conditions under which they lived can be charged against the men, not so their women at any rate. In the days when the wearing of home-made apparel was the common lot of the early settlers, the French Canadian women were noted for their domestic achievements, and that not only in respect to articles needed for their own household use but for marketing also. This was especially true in regard to two products, the braid for straw hats, and knitted woollen goods, especially mitts and socks. Both of these industries were prosecuted vigourously for many years in the early history of the parish, and even until very recent times. So late as 1881, the products of these two industries brought the French Canadian women of the country, an income estimated at $95,000 annually. It is on record that in 1880 the city of Detroit paid the women of Essex, $80,000 for straw braid alone. This braid was also exported east, and to South America and Mexico, where in both places it found a ready sale. These profitable sidelines continued unabated among these pioneer women until the advent of machinery and merging of capital and labour in the hands of large corporations, drove home-made productions out of the market.

The subsequent history of this settlement after the exodus of 1796, became merged in that of the whole country. This event added very few French Canadians to the ranks of their fellow countrymen on the Canadian side of the river. Apart from government appointees, none of them changed their place of residence. Whether the government of the country was associated or disassociated from the motherland of Great Britain was no concern of theirs. Later, they were to have settled convictions on these matters, but this was through the influence of subsequent additions to the population from Old France. When Republican France began to persecute the priests and destroy the property of the Roman Catholic church, many of them turned their faces towards Canada, where the Quebec Act of 1774, guaranteed them tolerance for their race, language and religion. Driven from their ancestral homes and their religious shrines by the persecutions of the devotees of republicanism, these were not likely to be found espousing enthusiastically republicanism anywhere, even in the United States of America. Representatives of this enforced emigration took up residence in the county of Essex, and with the three hundred and fifty hitherto settled here, gave to the Detroit river district a very pronounced French Canadian element in its subsequent population. During the years since, they have maintained intact their language, religion and race. They have their own schools and churches where possible, and favorite localities for settlement. Apart from that, it would be difficult to differentiate them from British Canadians, all alike exemplifying the virtues of industrial habits and law-abiding citizenship, and both striving equally to establish on these foundations, a prosperous and increasing community in this well-favoured section of the British dominions in Canada.

CHAPTER XIV.

The Pioneer British Settlement

Called the "New Settlement" to Distinguish it from
The Old or French Settlement

FOR more than twenty years after the defeat of the French in Canada, and the surrender of the Detroit river district to the British, the British Government made no attempt to establish British settlement in the district lying on the east side of the Detroit river. The French settlement on that side, by reason of Montreal Point being set apart as an Indian reservation, was separated into two divisions. The one on the south of the reservation, and extending in that direction to the Canard river, was named Petite Cote. The one on the north of the reservation we have been pleased to name Assumption, after the name of their parish. These three, Assumption, Petite Cote, and the Reservation occupied the whole frontage as far south as the Canard. From there to the mouth of the river, there was no settlement of any kind until after the close of the Revolutionary War. The first steps leading to the location of British settlers in the district, were undertaken, not by the Government, but by the prospective settlers themselves. Government took no direct action in the matter until 1788, when the establishment of a Land Board for the district of Hesse was made by Proclamation of Lord Dorchester.

167

The tardiness of the Government in making preparatory arrangements for the settlement of the district, was due to the many adjustments arising from the war, which required time for their satisfactory settlement. No well-defined policy of Land Settlement could be effected until these post-war problems had time to solve themselves. Two irregularities occurred on account of this delay. The first of these arose from the aggressive enterprise of individuals seeking locations.

Malden Purchased Privately from the Indians

Taking upon themselves the prerogatives of the Crown, these entered, on their own initiative, into an arrangement with the Indians for the surrender of the tract of land of which they were anxious to obtain ownership. This practice was extensively followed on the Michigan side of the river, "beyond all bounds of reason, in so much, that except the Hurons, there is not a Nation in that neighbourhood that has any property remaining." The most illustrious example of this method of anticipating Government action which we have had on the Canadian side of the river, was in the case of that block of land lying east at the mouth of the Detroit river and on the north shore of Lake Erie, in area about seven miles square.

On June the seventh, 1784, Alexander McKee, William Caldwell, Charles McCormick, Robin Eurphlect, Anthony St. Martin, Matthew Elliott, Henry Bird, Thomas McKee and Simon Girty, taking opportunity by the forelock, entered into an agreement on their own account with the Indian chiefs for the surrender of their claim to them of this valuable tract of land, although another claimant, Mr. Schiefflein, averred that he had a previous arrangement to theirs for the surrender of it to him. A properly ordered system would require that the Indians should surrender their claim in the first place to the Government, and after that, the Crown apportion to each individual his appointed share. In this case, the Crown and its representatives, were dispensed with altogether and arrangements made first among themselves and then with the Indians.

If this system of dealing with land settlement were permitted to become a settled practise, it would open the way to all kinds of abuses and irregularities. The Government, rightly, took the position that possession of the land from the Indian would have to be secured in the first place by the Crown, and then by the individual from the Government. In this instance, the men interested were all ex-soldiers from the Revolutionary War. They had rendered signal service in behalf of the British cause in that war, and were therefore entitled to a special claim upon the Government's generosity. Governor Haldimand, representing the Crown, because of this service and not that he was in accord with their methods, gave his sanction to their bargain making, and promised confirmation of the surrender by the Crown, and, in the meantime, until such confirmation was obtained, gave them permission to make clearances and erect buildings on the land which, by this method, they had obtained. The imprudence of sanctioning such unauthorized undertakings was emphasized later, when it was found that the Government needed a portion of this block of land for a military post and townsite, when the Jay Treaty made the evacuation of Detroit imperative.

The site selected was found to be a part of the holding set apart for Captain Henry Bird, who in the meantime had made a clearance of two hundred acres upon it, erected three dwelling houses, and planted a large orchard, all at any outlay of £1200. The expropriation was made at a time when the owner was on a visit to England, who therefore knew nothing concerning it until after it was completed. After his death, which occurred while he was on a military expedition to Egypt in 1801, some members of his family filed a claim for compensation from the Government, but his heirs received no redress, it being given out that this was a part of the agreement on the basis of which the transaction with the Indians was legalized by the Government.

First Locatees, Squatters

A second undesirable feature forced upon the first settlers by the tardiness of the government was the necessity of locating on the land before it was surveyed into townships or farm lots. In this way, the locatees got what was called a squatter's claim upon the land, but when the survey was later made, it was found very difficult to carry out a properly laid out plan of survey and maintain intact for each of these, the holdings which they had staked out for themselves and the improvements which in the meantime they had made upon their selections. "It is impossible," reported Patrick McNiff, the surveyor of the Lake Erie frontage, "to comply with the general plan of survey without injuring many of the inhabit- ants in their improvements." The first surveys of the district, consequently, had to be made to adjust themselves to these squatters' claims and in such a way as to preserve for them whatever of improvements they had made upon them, and on the instalment plan, without any well defined policy covering the whole territory, such as later was practised in the organis- ation of new counties.

It will be seen from this that if the intending settler was to be relieved from the necessity of establishing himself on a squatter's claim, some kind of government supervision ought to have been arranged prior to 1784. This need, Lord Dorchester met in 1788, when by proclamation he established a Land Board, to comprise five prominent men of the community, to deal with all questions relative to land settlement. This Board met at stated intervals, made arrangements for the survey of tracts of land being opened out for settlement, and apportioned to each applicant as the circumstances warranted.

As this Board did not begin work until September fourth, 1789, there was a period of five years from the arrival of the first settlers preceding this date without any government machinery established to aid them to select and to obtain possession of suitable locations, and twenty five years, a full quarter of a century, since the district had come into British possession. The circumstances of the times, however, were compelling attention to the question of land settlement and the development of the district agriculturally. The British Crown had ceded all its possessions in North America, save Canada, including Detroit and the territory west of the river. The close of hostilities discovered a great number of people domiciled in the United States, who wished to retain their allegiance to the British Crown in preference to Republicanism. Commencing in 1784, these made their way in increasing numbers to Canada, and many of them to Detroit, having in view settlement on the land of the Canadian side of the Detroit river district.

Loyalists and Pacificists Arrive from the United States

There were two classes of these intending settlers from

the United States, loyalists and pacificists. The loyalists were divided into two groups; first, disbanded soldiers of the British Regulars, and Hessian troops who fought in the American Revolutionary war; and, second, American citizens, who, fighting on the side of continued British connection, found themselves at the close of hostilities, the victims of persecutions on the part of the successful Republicans, and those of them with possessions, their property holdings taken from them. These former American citizens are the only class who have a right to the title of Loyalists, or United Empire Loyalists, as they were later designated.

There was, however, a fourth group among later immigrants, who styled themselves loyalists. To genuine loyalists, as well as to disbanded soldiers, the government had promised generous treatment. Was it because of these generosities, or the popularity which became attached to the name, at any rate, many who had neither handled a British gun nor faced a Republican soldier, adopted the title and sought to obtain land holdings on the strength of their pretensions. So successful were these in their applications for land, that at one time, the country was in danger of becoming a nation of these pseudo-loyalists, or pretenders. It would be well for those Canadian citizens of to-day, claiming loyalist descent, to examine carefully to see to which of these two classes of loyalists their progenitors belonged.

The pacificists, another class of prospective immigrants from the United States, were mostly from Pennsylvania, and of German descent. They were locally known by the name of 'Dutch Tories'. These took no part in the Revolutionary War, but because of their nationality, and the friendly relationship existing at that period of their history between Great Britain and Germany, were placed under the same odium as if they had. From these groups the first British settlements in the district were established.

The Disbanded Soldiers of Butler's Rangers

The first of these to arrive with a view to settlement, if we leave out of account such men as Alexander McKee, Matthew Elliott and Simon Girty, who had reached the Detroit before the close of the war, were representatives of the army unit known as Butler's Rangers. Disbanded at Niagara, these came to the Detroit district under the leadership of Captain William Caldwell, an officer of their unit. To him was given authority to treat with the Indians for a surrender of a portion of the district adjacent to Lake Erie, east from the tract exempted by Alexander McKee and his associates. These locatees became the first British settlement of the Canadian Detroit river district, and was known in the pioneer days of Essex as the New Settlement, to distinguish it from the French, or Old Settlement, on the Detroit river.

Captain Caldwell, a native of Ireland, was an officer in the British Regulars operating in the south during the first stages of the American Revolutionary war. Subsequently he was transferred to Butler's Rangers, a cavalry unit comprised of American citizens only, and organized for scout duty. In the continuance of hostilities, as the strength of this unit became depreciated through losses in killed, wounded, and prisoners of war, soldiers from the British Regulars were supplied to keep them up to strength, among whom J. G. Simcoe, the first Governor of the province of Canada, was an illustrious example. In this way, the original and strictly American identity of the unit was in a measure lost.

The undertaking to establish these ex-soldiers in one settlement and in that particular place, was not done from a wholly disinterested motive, having regard to their future well-being only. "As I look upon the Settlement mentioned in this letter to be in some degree a military one, in so much that it is to be composed of persons who have served in the course of the war, together, and considered by the Indians as connected with them for their mutual strength and benefit, you will be very particular in not permitting little traders and interested persons from creeping into it, and admit only those persons whose services and undisputed attachment to Government shall recommend them to the principal persons of the Settlement."*

The aim, therefore, from the Government viewpoint, was in the main a military one. The establishment of a defensive force for the protection of British possessions in the district, was deemed essential at that time in view of the atmosphere of the day and the cession of the town and territory, which had been agreed upon, west of the Detroit river.

Progress of Settlement Under Captain Caldwell

The experiment, though on the whole quite successful, was not unattended with difficulties and disappointments, which, with a more alert and efficient government supervision might have been avoided. Captain Caldwell was an effective soldier and an efficient army officer, and he has to his credit much valuable service performed for the development of the district in the pioneer days of its settlement. But two incidents in his career, as a pioneer of the Lake Erie settlement, affected his reputation injuriously. The first of these

---

* Major Robert Matthews to Sir John Johnson, Quebec, 14th August, 1784.

may be attributed to imprudence, and the second, probably, to carelessness. The former has been already mentioned, his participation with the Indians in their resisting battle against the advance of General Wayne's army at the Miamis Rapids. Captain Caldwell was highly regarded by the Indians and he wielded a great influence over them. His presence with a small group of the Canadian militia in that battle brought him no loss of prestige among them, but as a citizen of British Canada, it was an act, which with others, came dangerously near involving Great Britain in another war with the American Republic.

The other incident had to do with the carrying out of the arrangements entered into with the disbanded Rangers, by the Government, when settling them on the lands apportioned for them on the north shore of Lake Erie. These arrangements not only included a land grant but supplies—tools, farm implements and provisions—all of which were essential to the maintenance of these men and their families until they would have made sufficient clearance on their land to become in some measure self-supporting. The individual locatees did not receive these supplies, in consequence of which many did not settle on their land, nor perform the settlement duties required of them. Because of their failure to comply with these requirements, their grants became forfeited and were given to others less deserving of recognition by the Government. When, then, the Land Board began to function, and complaint was made concerning this matter, it was ascertained, after enquiry was instituted, that these supplies had been delivered by the Government, and placed under the care of Captain Caldwell as leader of the Settlement, but the reason for his failure to apportion to each individual Ranger his appointed share, has not been placed on record. In the meantime, and as a consequence of this failure, many of the Rangers moved to the United States, a loss to the defensive forces of the district which the country could ill afford. Others of them accepted locations on the Thames river in lieu of that which had been taken from them in the New Settlement.

Notwithstanding these miscarriages, and although not now wholly comprised of Rangers, by 1790, the New Settlement had stretched itself eastward for fifteen continuous miles from its starting point four miles east from the Detroit river. When it was surveyed, it was made to comprise one hundred and nine lots of about two hundred acres each. That year, the name was changed to 'Two Connected Townships' in the New Settlement, Lake Erie. A town plot was reserved, the first to be laid out in Canada west of Niagara, to which the name of Colchester was given. When the final survey was made, the Settlement was divided into two townships, Colchester and Gosfield, and from that time forth its development went on uninterruptedly, becoming one of the first of British settlements in importance as it was the first in the order of time. *

Thames River Settlement

At the same time as the New Settlement was getting its quota of settlers for the Lake Erie shore, a similar movement was under way on the banks of the Thames river. Detroit, notwithstanding that it was slated by Old Country treaty-makers to become a possession of the American Republic, was still occupied by the British merchants of the Detroit river district, who made it the centre of their trade with

---

*George F. Macdonald. Essex Historical Society Reports. Volume III.

JAMES BABY
Founder of the Baby Mansion, Sandwich

W. J. BEASLEY, M. D., SANDWICH, ONT.

Dr. Beasley, the occupant and owner of the historic
Baby Mansion, Sandwich, (see page 185), has shown great
interest in the preservation of the pioneer features of that
edifice which he utilizes for a dwelling and professional
office. He also has shown great interest in the compiling
of the story of the Canadian Detroit river region, and is
a life member of the Algonquin Association, called into
existence for the preservation of things historical in the
district.

neighbouring Indian tribes and villages. It was still the seat of the representatives of the British Government, with its garrison soldiers and its representatives of law and order for the district. And, it was also the headquarters for those immigrants coming into the country with a view to settlement in undisputed Canadian territory. As long as it continued to be the gathering place of these intending settlers, the opening up of the district had to be of necessity from the west, eastward.

The pioneer settler chose his location on the shore of a lake or on the bank of a river. He had of necessity to do this, since these were as yet the only highways in the country. By canoe, batteau, or sail boat, the pioneer transported his goods and there was no other method available until roads should be built. But in addition to the facilities for travel and the freightage of goods which these waterways supplied, the first settlers chose their frontages for settlement because of the flat nature of the country and the necessity of having some means of draining the land of the surplus water with which it was flooded every Spring of the year. The highlands on the banks of a stream or the shore of some lake provided these natural facilities. When the whole frontage on the east side of the Detroit river had been pre-empted, there was nothing for the other incoming pioneers to do but to go further afield, if they desired to continue at all as citizens of the British Empire. Thus while the Rangers were directing their attention to the north shore of Lake Erie, other land-seekers were exploring and locating on both sides of the Thames river.

From its mouth at Lake St. Clair, and eight miles eastward, the Thames river passed through a section of territory, which though exceedingly fertile, was too low to attract any pioneer immigrant risking a settlement there. But from that point, this geographical feature so rapidly improved that a continuous line of settlers' huts occupied both banks of the river eastward to the lower Forks, the site afterwards chosen for the present city of Chatham, and this before the Land Board at Detroit had issued its order for the first survey of the district into farm lots.

The Selkirk Colony

Further north, at the mouth of the Sydenham river, there was established in 1804, the ill-fated Selkirk colony.   Lord Thomas Douglas, of Argyleshire, Scotland, with the most humane of intentions, brought out one hundred and eleven of his tenants, with a view to establishing an estate in that section of the country patterned somewhat after Old Country methods. The experiment was a failure, due chiefly to the low nature of the locality chosen for their settlement.   Forty-seven of them died the first season, from exposure and the effects of malarial fever.   Those who survived, were compelled to seek other locations for settlement, flooded out by the inundation of the estate by a rise in the level of the waters of the lake and river.

Another settlement, which figured prominently in the War of the American Invasion, was a thriving Indian Mission established by Moravian Brethren, on the north bank of the Thames, at the easterly limits of the Western District.

These various settlements comprise the first endeavours of the Western District towards the realization of that progress which has since been so successfully achieved.

CHAPTER XV.

Sandwich

The Establishment of a Substitute Capital for the Ceded Town of Detroit

IF a writer were to attempt to set forth the various preceding causes which led to the founding of Sandwich as an urban centre on the Canadian side of the Detroit river, it would take him far afield. Great political and epoch-making events would be found occupying a causal place in its establishment and origin—French aspiration in founding a colony in North America; British aggression and warfare extending its North American boundaries at the expense of their French neighbours; and, more immediately, the necessity to make an exodus forced upon the loyal subjects of Britain residing in Detroit and Michigan, as a consequence of the cession of Canadian territory to the United States at the close of the Revolutionary war—these constitute three of the major causes giving rise to its existence.

For both the French and the British nations, at the commencement of their history in this area, Detroit held the status, both of an important fur-trading post, and the capital and judicial seat of the district. In the treaty of peace of 1783, and this again ratified by the Jay treaty of 1794, the territory on the right bank of the river, including the judicial capital, was assigned to the United States. By these treaties Canada lost Detroit but gained Sandwich. Certain of the fur-traders who looked to the Indians and the North-east territory for their trade, certain of the Canadians who owned cleared land on the east bank of the river, certain of the official staff —judge and lawyers, sheriff and constables, these ferried themselves across the river and made selection of Sandwich as their chosen site for a future city. This was in 1796.

In the founding of Sandwich, there was no impatient haste. The treaty of peace of 1783, they considered this for eleven years and the Jay treaty for two years more, before they took the final steps to effect an exodus. They were attached to Detroit by many endearing associations. To the French Canadians, it was their first and foremost western colony. To the British, it was endeared by the memory of Major Gladwin and his handful of garrison soldiers, resisting successfully the besieging tactics of the combined forces of all the North-West Indian tribes, under Pontiac, an incident of persevering courage without parallel in British colonial history. They left the one place to found the other, not from choice, but, of necessity.

The Municipal and Judicial Capital of the Western District

One hundred and thirty two years have elapsed since that event. What have the years since done for Sandwich? Up until the year 1850, the history of Sandwich is the history of the Western District. Whatever of importance it had during these years, it received it from the fact of its being its judicial and municipal seat. It supplied a jail for all the wrong-doers and the law-breakers of three counties. It supplied the Judge to pronounce sentence upon them. To execute his just judgements, it supplied a sheriff, who happened to be also, an English church clergyman. It supplied also a place of meeting and houses of entertainment for the district councillors, convening at times from these three counties. Take away these—the jail, the courthouse, the registry office, the municipal meeting house and the hotels—from pioneer Sandwich, and you take away Sandwich.

As a judicial seat and a municipal capital for these three counties, it has to be said, however, that it did not give universal satisfaction. This arose from two causes. The first of these was its geographical situation as the farthest distance possible west from the homes of the district representatives. The second was, the condition of the roads, or rather, the lack of roads, at that period in the country's history. It meant for these representatives, the choice of ploughing through mud on horseback or the alternative of making the journey on foot. Seventy miles of a walk for George Young of Harwich, the prince among the district councillors; seventy four for James Smith of Camden, the dean among the Wardens, and one hundred and thirty for Warden Hyde of Plympton, the pioneer representative of Lambton—these were requirements which could not be evaded at certain seasons of the year, if they would attend the courts or the municipal meetings of the District.

Discontent with this one-sided situation of the District capital was steadily on the increase. The first person to give it executive expression was Robert S. Woods, at that time a young lawyer of Chatham, though a native son of Sandwich. In his professional capacity, he supported a petition before the assembled parliament at Kingston in 1847, and succeeded in getting Kent County established as a new district with Chatham as its capital. This left Essex and Lambton associated together as the Western District, with Sandwich still the capital, an arrangement convenient for Kent, but affording no measure of relief to the representatives of Lambton. This new creation was, however, short-lived. A Municipal Act was passed in 1849 which abolished districts, and set local government on an entirely new footing. Counties and county councils, townships and township councils, were established instead, a system of government which became effective on the first day of January, 1850. Sandwich entered at that date on its new status as a County Town, a dignity which it still possesses.

Sandwich as a Commercial and Industrial Centre

It might be supposed that Sandwich, founded so early in the history of the Canadian district, ought to have arisen also in prominence as a trading, a commercial and industrial centre. For thirty years, it held undisputed possession of this field of urban activity, during which period it made some progress, but nothing commensurate with its opportunity or the needs of the surrounding country. With the establishment of a stage-coach village at Windsor in 1828 (and subsequently, the coming of the railway there) Windsor, and not Sandwich, became the chief centre of trade on the east or Canadian side of the river. It may not be without interest for us here to quote a comparison of growth written by a reliable historian concerning these two neighbouring urban centres, during the first few decades of their respective histories.

"The county town and judicial seat of the County of Essex is beautifully situated in the midst of a fine and well settled agricultural county, on the Detroit River, about two miles below Windsor, and nine miles below Lake St. Clair. This is one of the oldest towns in the Western section of Canada,

having been the district town of the Western District, where
all the judicial business of the three Counties, Essex, Kent
and Lambton, was transacted, and one would be naturally
disposed to inquire why it did not keep pace with other towns
and villages throughout the country—having now only a
population of about one thousand, whereas the population of
its rival town, Windsor, numbers about four thousand five
hundred. The reason, no doubt, for the more rapid increase
of population in the latter place—and, consequently, its
greater increase in wealth and general importance, was the
fact of its having more immediate connection with the City
of Detroit upon the opposite shore, and also its being upon the
line   of   the   Great   Western   Railway.—Notwithstanding
Sandwich has her attractions, being surrounded by highly
improved and beautiful farms, among which might be
mentioned the Prince estate, owned now by Charles Prince,
Esq., a son of old Colonel Prince, formerly member of
Parliament for the Western District. Upon this place there
is an orchard extending over twenty acres, containing the
choicest fruits of every description. The farm in all contains
about 300 acres, in which there are beautiful oak groves,
evincing remarkable taste.

Several newspapers were established in Sandwich, but
through some means or other ceased to exist. In 1846 a
paper was published, called the Western Standard; at that
time the town contained a population of 450.

It contains at present one Grist and Carding Mill, two
Tanneries, one Sash, Door and Blind Factory, one Saw Mill,
one Foundry, a Brick Yard, Shingle factory, two Wagon Shops,
Bakery, one Gunsmith, seven general stores, four groceries,
five Hotels, five Saloons, one Boot and Shoe Factory, two Pot-

ash Factories, one Brewery, five Churches, several Schools." *

It will be seen from the above enumeration of the industries and business places established at Sandwich, that there was not a sufficient number of them to give any appreciable prominence to the place. Of the thirty seven enumerated, it is doubtful if all of them provided employment for more than a hundred people including their owners. They were of sufficient number and variety to lift the place to the status of a good rural village, supply the needs of the surrounding district of farmers, but outside of that it occupied, at that time, no great place in the industrial and mercantile activities of the country.

The tardiness which was so conspicuous a feature of its growth in the first century of its history continued on in the commencement of the second also. It was not until within the period of the last decade, that any marked improvement occurred. At the end of one hundred and twenty years of history, its total population was rated at three thousand, but, commencing with that decade, a rapid change took place. In the first five years of the decade, it had increased to five thousand in population, and in the second half, to over ten thousand, thus more than trebling its size in the last ten years. But this growth is in respect to its residential areas. Whether this rapid growth in its residential population will eventually result into its development into a great commercial and industrial centre y .t remains to be seen.

Sandwich as a Place of Historic Interest

But apart from any material progress made by Sandwich the place is not without interest to persons who wish to

---

* Atlas—MacDonald, 1882.

discover connecting links of the present with the pioneer past. That it was selected as the camping ground of Major Rogers, when he came to receive the surrender of Detroit from the French; that it was chosen as the judicial seat and district capital, when Detroit, with the western territory attached to it, was handed over to the United States; that General Hull bivouacked within its limits in his first efforts at the invasion of Canada—these, and other incidents of the past, have had their influence in making it a place of peculiar historic interest.

There is retained, and in a good state of preservation, a monument of this past, in a connecting link which takes us back to the beginning days of Upper Canadian history. This is a building locally known as the 'Baby Mansion,' erected on the bank and facing the waters of the Detroit river. Once used as a fur-trading house, it is now occupied as a residence and owned by W. J. Beasley, a physician of Sandwich. It exhibits with fidelity the colonial style of architecture, and in its situation there are not wanting elements which lift it to the category of the picturesque and stable in appearance.

In its associations, this mansion is symbolic of French Canadian loyalty and patriotism. James Baby, the original owner, was a truly representative French Canadian citizen. He was born at Detroit in 1762, at which time it was a British possession. He was, therefore, a Frenchman by blood, but a British subject by birth. He afterwards became a British subject by choice. At a time when opportunity was given him to remain in Detroit and become a citizen of the United States, he chose instead to abandon his possessions in Michigan and Indiana, and to proceed with others to the British side of the river in order that he might continue a loyal subject of the flag under which he was born. His father Jacques Duperon Baby, was a scion of an ancient and noble family of France. When he was a young man, though settled first in Lower Canada he moved to the Detroit colony, where he was engaged in the cultivation of lands and in fur-trading at the time when French Canada passed over into the hands of the British.

James Baby was educated at the Roman Catholic seminary in Quebec. On the completion of his education he took an extensive trip through Europe, after which he returned home, and joined his father in his business of fur-trading and farming. This was shortly after the close of the American Revolutionary war in 1783. In Sandwich, as in Michigan, he followed the double occupation of cultivating his lands while at the same time he acted as agent for the North-West Company trading in furs. There was born to him in this mansion, a family of five sons and one daughter, all of whom, when quite young, were left motherless. Though quite young when this misfortune befel his home, Mr. Baby never sought to repair it by a second marriage.

At a time when many of his countrymen were giving a very indifferent support to British connection, he became a loyal and trusted servant of the British Colonial Government. At the inauguration of Upper Canada as a separate province, he was appointed a member of the Executive and Legislative Council, both of which positions he continued to occupy until his death. He was also selected by Governor Simcoe for the position of County Lieutenant of Kent, an appointment which gave him the rank of a Commander in the militia in the Western District at the outbreak of the war of the American Invasion in 1812. He was taken prisoner at the Battle of the Thames, but General Harrison, who utilized his house for his headquarters, granted him his freedom, and he removed with his family to Lower Canada, where he remained during the period of the American occupation of Sandwich. When the war was over, he was appointed Inspector-General for the province, and moved to Toronto, where he continued to reside until his death in 1833. His son Charles became occupant after him of the 'mansion,' after which it passed into the hands of several owners before it came into the possession of Doctor Beasley.

The mansion is symbolic, also, of an epochal era in the history of North America, of the century which saw the passing out of the forests and the race of men to whom they provided a living. The ceiling of the wide spacious hall, separating the right from the left wing of the house, still holds the hook from which was suspended massive scales capable of weighing a ton of furs. Looking through the glasses of our constructive imagination, we can see drawn up on the banks of the Detroit the frail birch-bark canoes which brought their cargoes of furs from the far north. We can see the stoic denizens of the forests, sitting around this hall, smoking their pipes, the last generations of a passing era and a dying race of men. Wyandot, Ottawa, Pottawatamie, Ojibway, there is not a representative of any one of these to be found to-day on the banks of the Detroit. But this building stands where it stood more than one hundred and thirty years ago, venerated by the truly loyal for its associations, while at the same time it adds to the appearance and to the interest of the town of which it forms a very historic part.

Situated in the midst of surroundings which are both picturesque and peacable, the mansion ought to be also, an emblem of peace. Instead, it comes down to us as symbolic of war. When the peaceable trade in furs of the Eighteenth century gave place to the hostilities of the War of the American Invasion in the beginning of the Nineteenth, this building became the scene of several interesting episodes connected with the war. When General Brock assembled his little army on the Canadian shore in preparation for that memorable attack on the army of Hull and the fortress in which he had taken shelter, he chose Sandwich for his place of encampment and this house for his headquarters. Of special interest to us is the story of his return accompanied by Hull, the General of the surrendered American army, after he had completed his successful adventure. At the dock in front of this mansion, both Generals disembarked; both walked up together to the house, the one victorious and the other a prisoner of war. Within this house, both sat down together, and listened, as an officer of the Canadian militia, John Beverly, afterwards Sir John Beverly, Robinson read out to them the terms of surrender which he had prepared on the instructions of his superior officer. Both of these Generals had made proclamations. Both of them, no doubt, had anticipations of success. But the Fates decreed that Brock should dictate the terms of surrender, and Hull submit to, and sign them. The end, however, was not yet. While General Procter slept the last night of his sojourn on the Detroit, in the guest-chamber of this 'mansion,' he was to be succeeded in turn by General Harrison, returning victorious from the Battle of the Thames. As was the war, so was the house, keeping in step with military events and the disposal of their respective armies on the checker-board of circumstance and fate. With these and many other associations connected with its history, it stands a landmark in respect to which there should be no lack of appreciation on the part of those interested in historic truth.

Sandwich as a Seat of Learning

Passing from the annals of war to those of peace, we consider next the educational facilities provided by the town of Sandwich. In the course of its history, two commendable attempts have been undertaken to make of the place a seat of learning. General Simcoe started out the educational system of the province by establishing several grammar schools in imitation of Old Country conditions. Sandwich was selected as the site of one of these. For a number of years it served as the medium for the education of the well-to-do of the town. In the course of events, it gave place, of necessity, to the system of free schools later established, and now in vogue throughout the province.

A more successful effort, though it was after several failures were recorded, was the establishment of Assumption College by the Roman Catholic church, the first church to undertake mission work on the Canadian shore of the Detroit. Following the purposed method of this religious body, to educate in their own schools the children and the young people of their own denomination, a school was erected in Sandwich as early as 1855. For fifteen years, it ran a somewhat chequered career, but a more promising destiny opened out for it in 1870. On that date, a number of Basilian priests came to Sandwich to assume direction of Catholic education there, of which the founding and promotion of Assumption College was to be their chief concern. Rev. Father Dennis O'Connor was appointed its Superior. The school started out with an enrolment of fifty-eight scholars. With this as a beginning, the

College has steadily grown and prospered. At the celebration, in 1920, commemorating the fifty years in which it had been in charge of the Basilian fathers, the College reported an enrolment of three hundred and fifty students, fifty per cent of these from Ontario, the remainder from Michigan and other states of the Union. To the original building erected in 1855, new buildings have been added from time to time, until there is now an imposing group of which the town and the Catholic church are justly proud.

The Jubilee year of 1920 saw another step forward in progress—the affiliation of Assumption with the Western University of London, Ontario. Students may now secure from this College at Sandwich, by reason of this affiliation, a University degree, and at the same time enjoy all the advantages given by a staff who make training in "Virtue and Discipline" as well as "Knowledge," the basic principles governing instruction.

While the commendable work of the Basilian fathers in founding Assumption College, now practically a University, is greatly appraised, the question forced to the forefront at the present time is, whether or not Sandwich will be able to supply in the future the environment suited to a seat of learning. Up until the present time Assumption provided a quiet resort where students might meet and think undisturbed by the clash and clamour of the activities of this industrial era which we are all passing through so noisily. A great undertaking however, which may make a complete change in the future of Sandwich, is now being achieved. This is the spanning of the Detroit river with an international steel bridge. What changes the establishment of this enterprise will effect on the prospective outlook of the town, the future alone can tell

CHAPTER XVI.

The Beginning Years of Amherstburg

A Little Town of Twenty Houses
Becomes a Provincial Naval Station

DURING the first fifteen months of the War of the American invasion, there was no place in Canada which was more in the thought of the military leaders of the United States, than the military post at Amherstburg, then named Fort Malden. This was due to the fact that the events of that period effected a complete somersault in American expectations.

The year 1812 inaugurated an unhappy set of circumstances for the infant settlements of Upper Canada West. In that year, the feud created by the Revolutionary war found a new outlet and expression in a second war, ostensibly against Great Britain, but in actual reality, against Canada. The hand of the United States Government was forced by the Western States, notably Pennsylvania and Virginia, who desired war among other reasons that they might invade and capture Canada and incorporate it into the territory of the States of the Union. They had no hesitating doubts as to the issue. They would strike quickly and strongly and there could be only one result.

They based their expectations on two disabilities under which Great Britain then labored in respect to the protection

191

of Canada. The first was, her pre-occupation. She was engaged, and that to the full extent of her capabilities, in her old-time occupation, a war with France. The second was the unprotected condition of Canada. At one end of her far-flung frontier was Quebec; at the other end, Amherstburg. Between these two were many miles and only four garrison posts, and each of these manned by a small handful of soldiers.

Quebec was the most strategic point to strike at first. If the fall of that fortress were achieved, the other posts Kingston, York, St. George, Erie and Amherstburg, cut off from Great Britain, the source of their supplies, must of necessity, surrender also. But instead of Quebec receiving the first place of attention, Amherstburg was accorded that distinction.

Amherstburg was founded with a view to making it the Gibraltar of the Canadian Detroit. With the passing of the Jay Treaty, it became the substitute of Fort Detroit as the place of habitation for the British garrison stationed to keep peace with the Indians in this district. As the treaty was passed with a view to avert war between Great Britain and the United States, and since in the Old Country, it was deemed to have accomplished this end, the original intentions and plans for fortifications were much modified. In the first draft, it was intended to erect two blockhouses, one on either end of Bois Blanc island, with a strong palisaded fortress on the mainland midway between the two. These plans, however were carried out only in part. An era of long peace was expected by the authorities residing in Great Britain, who in this matter, as in many others, exhibited a woful lack of knowledge of the country and its existing conditions in relationship to the frontier settlements of the United States.

MAJOR JOHN RICHARDSON

Upper Canada's first novelist and the premier historian of the war of 1812, was born at Queenston, October 4th, 1796, the oldest son of Dr. Richardson, a surgeon attached to the British army doing garrison duty in Canada, and Madeleine Askin, a daughter of the pioneer, the well-known John Askin of Detroit and later of Strabane, on the Canadian shore. He commenced his career as a school boy in Amherstburg, which he gave up at fifteen years of age, to join the company of the 41st regiment stationed at Fort Malden, on the occasion of the outbreak of the War of the American Invasion, 1812.

After the war, he devoted his life to writing, which brought him a very precarious livelihood. He died in extreme poverty, if not in absolute want, May 12th, 1852, in the city of New York, and was buried somewhere outside of the city, but the place unknown to his friends or posterity.

MAJOR JOHN RICHARDSON

"I had first breathed the breath of life near the then almost isolated
Falls of Niagara—the loud roaring of whose cataract had, perhaps, been
the earnest of the storms—and they have been many—which were to
assail my life after. My subsequent boyhood, up to the moment, when
at fifteen years of age I became a soldier, had been passed in a small
town (Amherstburg), one of the most remote, while, at the same time,
one of the most beautifully situated in Canada. I had always detested
school, and the days that were passed in it were to me days of suffering,
such as the boy alone can understand. With the reputation for some little
capacity, I had been oftener flogged than the greatest dunce in it,
perhaps as much from the caprice of my tutor as from any actual wrong
in myself—and this had so seared in my heart—giving me such a disgust
for Virgil, Horace, and Euclid, that I often meditated running away, and
certainly should have gratified the very laudable inclination, had I not
apprehended a severity from my father—a stern, unbending man, that
would have left me no room for exultation at my escape from my tutor.
It was, therefore, a day of rejoicing to me when the commencement of
hostilities on the part of the United States, and the unexpected appearance of a large body of their troops, proved the signal of the "break up"
of the school, or college, (for by the latter classical name was known
the long, low, narrow, stone building, with two apologies for wings
springing at right angles from the body), and my exchange of Caesar's
Commentaries for the King's Regulations and Dundas. The transition
was indeed glorious, and in my joy at the change which had been wrought
in my position, I felt disposed to bless the Americans for the bold step
they had taken."—Eight years in Canada, by Major Richardson.

A fortress capable of withstanding the assaults of a few hundred Indians was all that was adjudged necessary.

The selection of this place as a site for a strong fortress was assuredly well chosen. Protected on the one side by the Detroit river, on the other by Lake Erie, and with Bois Blanc island on the front of it, an aid to fortification, a better location for defensive purposes could hardly be chosen. It commanded the entrance of Lake Erie from the west and the Detroit river from the south and east. The Americans planned an attack upon it as the very first military stroke of the western campaign. Strange as it may appear, this place, selected as a spot for western defence, against which the Americans mobilized four successive armies with a view to the defeat or capture of the garrison stationed there, and the reduction of the Fort, such as it was, yet, notwithstanding all this, the town of Amherstburg has no such event to chronicle as part of its past history. The first army delayed its purpose too long. Their second and third armies were met on their own soil by the combined Canadian forces, made up of a few hundred British regulars, the Essex and Kent militias, the North West Fur Company's engages, and Indian warriors, notably the Hurons under their chief Roundhead of the Detroit river district, and after stubborn engagements in both cases, the Americans suffered terrible and slaughtering defeat. Their fourth army, six thousand strong, the first, a portion of it landed unhindered three miles east of Amherstburg, after the defeat of the Canadian squadron on Lake Erie, but when they arrived at the mouth of the Detroit, there was no fortress there. General Procter, when he abandoned the west destroyed the ramparts and removed their equipments, thus taking away all opportunity for its capture or reduction.

The fifteen months preceding this wanton destruction of the aims for which Fort Malden was called into existence, constitute the greatest period in the history of the locality. Next to Sandwich, Amherstburg is the oldest urban centre of western Ontario. The site of the present town figured in the first land transaction to take place in the district of Malden. It was part of a block of land seven miles square purchased by a group of private men directly from the Indians. This transaction was decidedly irregular, but the parties to the transaction were all persons who had rendered valuable service to the Crown during and after the war of the American Revolution. It was therefore made an exception, and later ratified by the Crown, each man in this way confirmed in his holding. In granting confirmation, stipulation was however made that a portion of the block facing the Detroit river would be required by the Government for a townsite. When the selection was made, it was found to be a part of the holding of Captain Henry Bird, a soldier of the British regulars, which later, they expropriated.

It may not be without interest to insert herewith a description illustrating the progress made by the town by an eye-witness, Isaac Weld, at the beginning of the Nineteenth century, about a dozen years preceding the outbreak of war:

Malden

Malden is a district of considerable extent, situated on the eastern side of the Detroit River, about eighteen miles below the town of Detroit. At the lower end of the district there are but few houses, and these stand very widely asunder; but at the upper end, bordering upon the river, and adjoining to the new British port that has been established since the evacuation of Detroit, a little town has been laid out, which already contains more than twenty houses, and is rapidly increasing. Hither several of the traders have removed who formally resided at Detroit. This little town has as yet received no particular name, neither has the new port, but they merely go under the name of the new British port and town near the island of Bois-Blanc, an island in the river near two miles in length, and half a mile in breadth, that lies opposite to Malden.

When the evacuation of Detroit was first talked of, the island was looked to as an eligible situation for the new port, and orders were sent to purchase it from the Indians, and to take possession of it in the name of his Britannic Majesty. Accordingly a party of troops went down for that purpose from Detroit; they erected a small block house on the northern extremity of it, and left a serjeant's guard there for its defence. Preparations were afterwards making for building a fort on it; but in the mean time a warm remonstrance against such proceedings came from the government of the United States, who insisted upon it that the island was not within the limits of the British dominions. The point, it was found, would admit of some dispute,and as it could not be determined immediately, the plan of building the fort was relinquished for the time. The block house on the island, however, still remains guarded, and possession will be kept of it until the matter in dispute be adjusted by the commissioners appointed, pursuant to the late treaty, for the purpose of determing the exact boundaries of the British dominions in this part of the Continent, which were by no means clearly ascertained by the treaty of peace between the States and Great Britain.

In this particular instance the dispute arises respecting the true meaning of certain words of the treaty. "The boundary line," it says, "is to run through the middle of Lake Erie until it arrives at the water communication between that lake and Lake Huron; thence along the middle of the said water communication. " The people of the States construe the middle of the water communication to be the middle of the most approved and most frequented channel of the river; we, on the contrary, construe it to be the middle of the river, provided there is a tolerable channel on each side. Now the island of Bois Blanc clearly lies between the middle of the river and the British mainland; but then the deepest and most approved channel for ships of burthen is between the island and the British shore. In our acceptation of the word, therefore, the island unquestionably belongs to us; in that of the people of the States, to them. It appears to me, that our claim in this instance is certainly the most just; for although the best and most commodious channel be on our side, yet the channel on the opposite side of the island is sufficiently deep to admit through it, with perfect safety, the largest of the vessels at present on the lakes, and indeed as large vessels as are deemed suitable for this navigation.

Plans for a fort on the main land, and for one on the island of Bois Blanc, have been drawn; but as only the fort will be erected, the building of it is postponed until is is determined to whom the island belongs; if within the British dominions, the fort will be erected on the island, as there is a still more advantageous position for one there than on the main land; in the mean time a large block house, capable of accomodating in every respect comfortably, one hundred men and officers, has been erected on the main land, around about four acres or more of ground have been reserved for his Majesty's use, in case the fort should not be built on the island.

A block house, which I have so frequently mentioned, is a building, whose walls are formed of thick square pieces, of timber. It is usually built two stories high, in which case the upper story is made to project about two or three feet beyond the walls of the lower one, and loop holes are left in the floor round the edge of it, so that if an attempt were made to storm the house, the garrison could fire directly down upon the heads of the assailants. Loopholes are left also in various parts of the walls, some of which are formed, as is the case at this new block house at Malden, of a size sufficient to admit a small cannon to be fired through them. The loopholes are furnished with large wooden stoppers or wedges, which in the winter season, when there is no danger of an attack, are put in, and the interstices closely caulked to guard against the cold; and indeed, to render the house warm, they are obliged to take no small pains in caulking the seams between the timber in every part. A block house, built on the most approved plan, is so constructed, that if one half of it were shot away, the other half would stand firm. Each piece of timber in the roof and walls is jointed in such a manner as to be rendered independent of the next piece to it; one wall is independent of the next wall, and the roof is in a great measure independent of all of them, so that if a piece of artillery were played upon the house, that bit of timber along against which the ball struck would be displaced, and every other one would remain uninjured. A block house is proof against the heaviest fire of musquetry. As these houses may be erected in a very short time, and as there is such an abundance of timber in every part of the country, wherewith to build them, they are met with in North

America at almost every military out-post, and indeed in almost every fortress throughout the country. There are several in the upper town of Quebec.

Amongst the scattered houses at the lower end of the district of Malden, there are several of a respectable appearance, and the farms adjoining to them are very considerable. The farm belonging to our friend, Captain E.—,under whose roof we tarry, conditions no less than two thousand acres. A very large part of it is cleared, and it is cultivated in a style which would not be thought meanly of even in England. His house, which is the best in the whole district, is agreeably situated, at the distance of about two hundred yards from the river; there is a full view of the river, and of the island of Bois Blanc, from the parlour windows, and the scene is continually enlivened by the number of Indian canoes that pass and repass before it. In front of the house there is a neat little lawn, paled in, and ornamented with clumps of trees, at the bottom of which, not far from the water, stands a large Indian wigwam, called the council house, in which the Indians are assembled whenever there are any affairs of importance to be transacted between them and the officers in the Indian department. Great numbers of these people come from the island of Bois Blanc, where no less than five hundred families of them are encamped, to visit us daily; and we in our turn go frequently to the island, to have an opportunity of observing their native manners and customs.

Our friend has told them, that we have crossed the big lake, the Atlantic, on purpose to come and see them. This circumstance has given them a very favourable opinion of us; they approve highly of the undertaking and say that we have employed our time to a good purpose.

CHAPTER XVII.

A Nation at War Against a Settlement

The War of The American Invasion, 1812

IN the march of circumstances, a period arrived in the history of the Western District, in which the armies of the newly created republic of America invaded the Detroit frontier, an event which, for a time, greatly retarded the progress of the settlement. The people of the American nation, especially that section of them inhabiting the Western States, were much dissatisfied with the findings of the Commission, appointed to determine the terms of peace, 1783. Since they did not obtain the "cession of the whole of Canada, including Nova Scotia," which Messrs. Franklin, Adams and Jay, the American representatives demanded, and in which the British representative, Mr. Richard Oswald, concurred, they determined to invade the country with a view to its subjugation and the incorporation of it into the United States as a part of the Union. This was not the reason which was given then, for the attack upon this country, nor is it the reason which is given to-day, by writers whose business it is to defend the past, not to truthfully record it. But, if the speeches of their leaders in promoting the war, and the subsequent action of those of them who obtained control of the Government, are true expressions of the sentiments of the nation, then we must conclude, that this, and this alone, was the inspiring motive which led to the invasion.

In respect to the Detroit River district, the war there became, as far as the Canadians were concerned, the attack of a nation against a settlement. In their plan of campaign, the Americans prepared three separate army units, each with its own objective. One army was destined for the east with a view to the reduction of Fort Quebec. Their second army was to strike centrally at Niagara; while their third, and the first on the ground, was intended to bring about the immediate capture of the whole of the Detroit River district, northward. This third was, of all their campaigns, the one which supplied the most dramatic incidents of the whole war, both in respect to their failures and successes. It was the one in which the pioneer settlers of this district were thrown back mainly on their own resources to provide defensive armies which were to withstand the onslaught of the American nation against them, in order to maintain in their own hands the determining of the destiny of their own country.

In the western states, where the desire to incorporate Canada into the Union was strongest, there was a prevailing sentiment that the capture of the Detroit River district would be very easy of accomplishment. They had not the least fear or misgivings in respect to the issue, and this assurance was the sentiment mainly responsible for the war. As a matter of fact, the majority vote, which committed their Congress to resolve themselves into an invading enemy nation against Canada, was supplied by two of these alone, Virginia and Pennsylvania. Apart from these two. the vote for and against was an even fifty percent for each side.* The armies mobilized for their western campaign were recruited from the western

*The total vote in the House of Representatives was, 79 for, and 49 against. The total vote in favor, in Pennsylvania was 16, and Virginia, 14.

states and commanded by western leaders. These were well-acquainted with the country and its resources for defence. Their hopes for a speedy victory and the early termination of the war were based on assured knowledge. It would appear, then, that their judgment was founded on rational grounds.

The Canadian settlement, against which they brought an army of twenty three hundred men, was defended by one garrison unit of British regulars—two hundred and fifty men of the Forty-first Regiment. This was a Regiment founded by Royal Warrant in 1719, and recruited from old soldiers of the Guards, Horse and Foot, at Chelsea hospital near London, at which time it was known as a Regiment of invalids. In 1798, the Regiment was in Ireland. In 1800, it embarked for Upper Canada and hence was on the ground when the war of the American Invasion took place twelve years later. They were destined to take a prominent part in the defence of the Detroit River district for the first fifteen months of this war, and to add to their colors the names of Canard, Maguaga, Fort Detroit, Frenchtown, Miami, Fort Meigs and Moraviantown as engagements in which they were to take a prominent part, in all of which, save the last, their reputation for successful undertakings, received enhancement. After the war was over, the Regiment left Canada for the Old Country, embarking from Quebec in 1815, but arriving at Ostend a month too late to be participants in the victory of Waterloo. At Amherstburg, they were under the command of Lieutenant-Colonel St. George, when the attack of the Americans upon the west was staged. He was superseded in the command of the Regiment on the arrival of Cononel Henry Procter, and appointed to the command of the First Brigade on the arrival of General Brock to undertake the attack upon Fort Detroit. This Brigade consisted of detachments of the Royal Newfoundland Fencibles, the Kent Militia, and the First and Second Regiments, Essex Militia.

The attack upon the western frontier stands by itself as a unique part of the whole war. The defence of the settlement was a distinctively Canadian undertaking, save for the leadership given by the two hundred and fifty British Regulars and their officers. Fortunately, preparations were not delayed until after the Americans had struck their first blow. There was not found amongst the fur-traders that credulity, in respect to the spirit and aspirations of the Americans, which was found amongst the eastern leaders. They knew well the intentions and purposes of the western states. When the army of General Hull was mobilized, Indians and others kept them constantly informed of its movements.

The Americans had two strong fortresses which they had obtained by cession from Canada in 1783, Fort Detroit and Fort Michilimackinac. Using these as a base, General Hull's army was marched into the Detroit region in order to be in readiness to strike a prompt and unlooked for blow at western Upper Canada, and affect its capture immediately following the declaration of war by the United States. It was a plan of campaign that had military prescience inspiring it, but it required two circumstances for its successful realization— first, an immediate and unexpected attack on the part of the American forces upon Fort Amherstburg at the mouth of the Detroit River, and Fort St. Joseph in Lake Huron; and, second, unpreparedness on the part of the Canadians to meet the attack with any power of resistance. Had these two factors been supplied, there is not the least doubt but that American expectations would have been fully realized. Their capture of the country would have been a matter of words, not deeds, as their western leaders ardently hoped and believed.

The first incidents of the war, however, were not staged by the Americans but by Canadians. The premier in importance as well as in time, was undertaken by a native-born Canadian, Lieutenant Frederic Rolette. He was an officer of the Provincial Navy, who had some time previously been appointed to the command of the brig, Hunter, with its naval crew of Newfoundland Fencibles and Canadian voyageurs. When war was declared, June 18th, this vessel was instructed to patrol the waters at the entrance of the Detroit River, in order to intercept any enemy vessels using the river to aid the garrison or army at Fort Detroit. The first to attempt the passage was a packet vessel, named Cayahoga, which disembarked from Miami for Detroit, having on board the baggage of the officers of General Hull's army, a number of sick soldiers, and military mail, including letters of correspondence passing between General Hull and the Secretary of War, revealing in detail their plan of campaign. This vessel Lieutenant Rolette took into possession on its arrival at the mouth of the river and towed it, with its valuable cargo, to the Canadian Naval port at Amherstburg. The value, for military purposes, of the capture of this correspondence, cannot be over-estimated. To this important achievement, he added a second, the capture of eleven American batteaux, laden with provisions. These, with their escort of soldiers, he towed in like manner to the harbor at Amherstburg. By these two premier incidents, he served notice on General Hull that his line of communication was in the possession, not of the invading Americans, but the defending Canadian forces.

What Lieutenant Rolette, with his crew of Canadian

voyageurs did on the water, Captain Elliott, the eldest son of Colonel Matthew Elliott, with a band of Indian warriors under Tecumseh, their Chief, duplicated on land. These two incidents were significant in that they indicated the ability of the Canadian forces to strike a telling blow against the American endeavour to subjugate the Canadian West and incorporate it, a part of the domain of the newly-created Republic. General Hull found his plans defeated, before he had reached the spot where he was to put them into effect. This Commander has had to suffer the consequences, not from circumstances which he had himself created, for which he would be justly to blame, but for the preparedness of the little band of Canadian defenders to create a set of circumstances to meet which the American Nation had not previously made provision. Colonel St. George, Commodore Hall, Lieutenant Frederic Rolette, Colonel Matthew Elliott, Robert Dickson, Thomas McKee were all on the alert before and while the American army was being mobilized in Ohio. They knew for what intent this army was being recruited. They were not affrighted by its size, nor did they deem themselves helpless. They did, what any wise leadership would have done under the circumstances. They gathered a defensive force from every available supply, and were preparing to meet that army, not cooped up in the Fort at Amherstburg, but in the open field of battle. They separated General Hull from his supplies and awaited in calm spirit, but with their rifles in their hands, the movements by means of which he would seek to extricate himself from the position in which their strategy had placed him.

I.

THOMAS BLIGH ST. GEORGE

Lieutenant-Colonel, Inspection Field Officer, Commander of Amherstburg Garrison Fort, 1812

His Efficient Preparatory Measures to Resist the Invasion of Canada, by the American First Army Under General Hull

IF you visited Amherstburg four years after the Exodus from Detroit took place, your attention would be directed, if you were observant, to two important features of this five-year old town—the one, the fort; the other, the shipyard. In the shipyard you would see a vessel being constructed, and if you enquired whether it was to be a merchant vessel only, or a gunboat, you would be answered, both. Even trading vessels carried a gun in those days, for whitemen were dealing with Indians, and of these Indians, some, who were friends to-day, would be enemies to-morrow. The whiteman took no chances in peace or war, but carried his fieldpiece, sometimes for defence and sometimes for offence. Thus the trading vessel furnished with a gun or guns, made provision for the protection of the trader's person and property.

When you looked in the direction of the Fort and saw soldiers drilling or making improvements on their garrison property, you would ask, "Is there a likelihood of war?" The pacificist, or the optimist, would answer you, "Why certainly not. We evacuated the post of Detroit peaceably. We gave up all the territory that both the treaty of 1783 and the Jay Treaty demanded as their price for peace. Only the Indian is now left with whom we might go to war, but he is friendly, we trading with him and he with us."

205

Yet, notwithstanding the many reasons which might be adduced in expectation of long-continued peace, it still remained that, up the river eighteen miles, an American garrison was occupying Fort Lernoult, and throwing hostile glances across to the other shore, while at the mouth of the same stream, a British garrison was laying plans for block-houses and for naval ships. Men were thinking of life and property in terms of guns and knives and tomahawks. The law of the land was still, "Not he that discovers, nor he that occupies, but he that can HOLD, shall be the owner thereof." British fought against British in 1779, and they might do so again, should an occasion arise to provoke it, and these occasions were not hard to discover when men were out in quest of them. Here, then, these sentinel soldiers drilled awaiting the coming of that occasion.

And not many years did they have to wait. The new century had run a course of twelve short years only, when arose the occasion, or what was thought an occasion. By a majority vote of thirty, in the House of Representatives, seventy nine for and forty nine against, and a majority of six in the Senate, nineteen for and thirteen against, the United States declared war against Great Britain, June 18th, 1812.

Few people to-day, having the well-being of the human race in their mind, seek to glorify war. Few seek to turn the mind of the young towards it, by picturing it as the alone place where bravery counts and where courage is shown. The world knows that war is a crime against humanity, a relic of barbarism, which brings a blessing to no people, but places a curse on many. The many are made to serve and die because a ruling few have not wisdom or grace enough to settle their differences amicably. "It delights in blood, and in fields strewn with carnage, in the tears of the widow and the plainings of the orphan perishing of want and disease."

In this instance, the War of the American Invasion was not of Canada's choosing. In the face of the Tenth Command of Mount Sinai, it was forced upon the country, in the first place, to satisfy the land-hunger of the people dwelling in the country south of us. Certain sections of our country was reputed among them, as being fertile land. Other parts of it was well known to be at that time a good hunting-ground for beaver and other wild animal pelts. This basic motive was supported by another, a political one. Inspired by their zeal for the Republican idea of government, they undertook the invasion of the country with the aim to have the whole of North America incorporated as one nation under one flag, and that the Stars and Stripes. A third motive, and probably a greater than either of the other two, might with truth be added as an influence inducing them to undertake war—the belligerent spirit of the age, fortified by feelings engendered by a previous war.

Across the ocean, on the continent of Europe, one of the greatest wars of all time was being waged by that prodigy of the times, Napoleon Bonaparte of Corsica. All the resources of Great Britain were being taxed to their utmost to thwart the ambitions of this upstart monarch.

The effects of war cannot be confined to those nations immediately participating in it. It affects also neutral nations. Causes of irritation will arise, and if any neutral nation be anxious to enter into conflict with either of the belligerents, it will not have far to go to seek an occasion. In this case, a too zealous search on the part of British seamen on board of American vessels, in search for deserters from the British navy, supplied the all-sufficient excuse for entering into a war with Great Britain, and thus provided an opportunity for the invasion and capture of Canada.

While the statesmen of Great Britain were debating the necessity for friendship with the Americans, and were passing enactments to remove existing causes of irritation, at the same time, in America, preparations were being made by the United States for their premeditated war against her. Fully one month before its declaration, an army of over two thousand in strength, was mobilized at Dayton, Ohio, and placed under the command of Brigadier-General Hull, who formerly had been a distinguished officer, in the Revolutionary War. All preparations to supply it with a full complement of equipment and provisions having been successfully achieved, this army started out on its mission the first day of June, 1812, with Detroit as its objective and the invasion and conquest of Canada, its ultimate aim.

The nation which is the aggressor, has a favourable advantage over the other, in that having purposed war, it can delay the declaration of it until the time best suiting its own convenience. This advantage, the Americans fully possessed. No preparations for this war were made by Great Britain, for it did not expect nor desire to enter into another era of hostile conflict with its former colonies. Instead, it did everything it could possibly do, to remove any irritating causes which might lead to it. Moreover, they were encouraged to think that there would be no war, by reason of the lack of unanimity among the United States people themselves. The eastern states were almost unanimous in opposing war, but the western states, which had their own reasons for demanding it, were in the majority, and they found a willing instrument for carrying

BLOCK HOUSE

BOIS BLANC ISLAND

Opposite Amherstburg is the beautiful island known as Bois Blanc, from the luxurious growth of white wood which originally covered it. It contains nearly three hundred acres of the finest land. The forest was cleared off it in 1837, to allow range for guns on the mainland, in case of insurgents effecting a landing on the island. Three block houses were built next year, for a number of years the island was garrisoned by a force of British troops. The government built a lighthouse on the southern point of the island in 1836. One of the old block houses still remains. The island, once the private property of Colonel Rankin of Windsor, is now owned by the Detroit and Windsor Ferry Company. The main channel runs between Bois Blanc and Amherstburg, and as it is very narrow here—more so than any point in its entire course—the season of navigation is continually enlivened by the many craft engaged in the lakes trade, which pass within literally less than a stone-throw off the mainland.

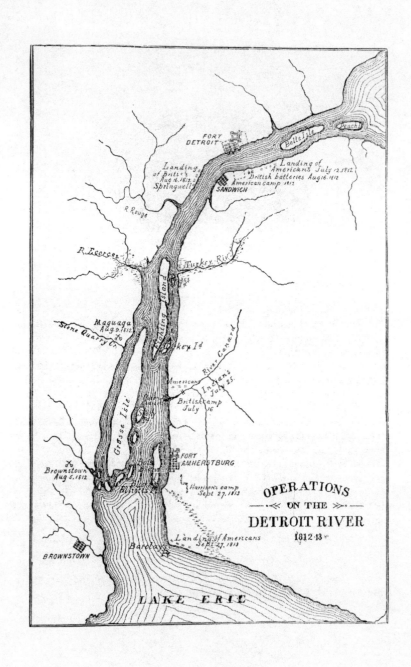

Operations on the Detroit River 1812-13

out their purposes in President Madison.

The premature mobilizing of so great an army, and its hurried march to the Detroit river district had as an objective, the reduction of Fort Malden by a surprise attack, to induce the Indians, by this gesture of strength, to ally themselves on the side of the Americans, or failing in this, to compel them to stand aside in a state of neutrality. Looking back on the century or more since, and considering the weak and sparsely settled district then called Canada, one wonders how a failure of the aims of these undertakings could have taken place. The failure of their western campaign is the most marvellous of all, for fail it certainly did, notwithstanding the seeming successes of September and October, 1813. When the struggle began, the Americans had three fortified posts, Detroit, Mackinac and Fort Dearborn, on the site where now is situated the city of Chicago. The Canadians had two, Fort Malden and Fort St. Joseph. When it closed two of these American forts had been captured and destroyed in the course of the struggle, Fort Detroit and Dearborn, while their third, Fort Mackinac, was captured by the British in the beginning of the struggle and held unsurrendered until its close. For this miscarriage of American expectations, the preparatory arrangements effected by Colonel St. George, must be accorded a measure of credit.

The first move on the checker board is oftentimes the deciding factor in the game. So also on the checker board of fate. General Hull was entrusted with the first movements on the American side, and Colonel St. George on the Canadian. If there was nothing brilliant in any of these preparatory movements staged by Colonel St. George, there was nothing, on the other hand, unwise. His first effort was to gauge the strength of the enemy, and then, ascertain, if possible, his objective. A reconnoitering party, of which himself was one, was sent to observe their landing on the Canadian shore, July 12th, 1812. Colonel St. George concluded, and that rightly, that the reduction of Amherstburg was their objective. Reducing this fort, and taking its handful of garrison defenders captive, there would be nothing remaining for General Hull to do but to march triumphantly eastward, and joining with the central expedition, at Niagara, repeat his successes there, and then both together continue their triumphant march to Montreal and Quebec. This was not the expectation of a wild and unreasoned enthusiasm. It was a cool judgment, based on unerring facts of favourable circumstances, the weakness of the Canadian defensive forces, the paucity of their resources and the strong element of indifference, if not even of disloyalty among its scattered inhabitants.

But Colonel St. George directed his attention, not to the resources which he did not possess but to those which he knew were available. If there were among the inhabitants, settlers who were indifferent or adverse to war, some of whom would desert their own militia forces to join the ranks of the enemy, as did Matthew Dolsen of the Thames river settlement, there were others who were differently minded. There were many among these settlers who were affectionately attached to the land of their birth, who prized its institutions, its laws, its government and its Crown as things above value, and for whose protection they deemed it worth while to sacrifice even life itself. These were cherishing the idea, with a hopeful forward look, that in a near-by day, a duplication of these institutions, laws and government would be set up in this land which they had now chosen to adopt as their own, and to which they had moved with their families. In the spirit of the lion robbed of its whelps, they were prepared to defend the children seated around these home firesides. There was, therefore, in prospect for Colonel St. George, a body of loyal Canadian militia, built on the foundations laid by Governor Simcoe, and while these might not be a formidable force, judged by their numbers and training and equipment, yet because of their spirit, they were still a factor which the invading army might find a strong obstacle in the way, preventing him from carrying out his projected plans of triumphant invasion. Colonel St. George's first step, after setting his own immediate house in order, was to call these men to the colors.

There was also another source of aid, and an important factor also, in building up a defensive force—the Indian. Inconstant, untractable, undependable, untrained in scientific warfare as he undoubtedly was, he yet could be made into a worth while factor in robbing the invader of his crown of triumph. There were men, alert to the gravity of the situation, long schooled in the method of handling the Indian, such as Robert Dickson, Matthew Elliott, William Caldwell, Thomas McKee, Lewis Crawford, John B. Askin and others, who had already begun to render invaluable service in calling to Canadian aid, warriors of tribes from the surrounding districts. Many Indians, of course, stood aloof for a time, their neutrality having been purchased by the Americans, but there were many besides these throughout these Lake Regions, who needed only wise leadership to enlist them in active support of the Canadian cause. That leadership was by no means lacking.

Next in importance to the enlistment of Indians domiciled

in Canada, was the enlistment of detachments of American Indians, who were induced to align themselves on the side of the British by Tecumseh, the chief of a small band of Shawnees. In the breasts of these Indians, there rankled the memory of burnt cornfields, destroyed villages, and homeless children under the generalship of Anthony Wayne in the region of the Miami river, and still more recently under the generalship of Harrison at Tippecanoe.

Thus with energy and activity, Colonel St. George strengthened his resources, knowing that valor alone could obtain no victory in the face of overwhelming numbers.

Movements towards increasing his strength, having thus been initiated, the plan of defence prepared by Colonel St. George, was to meet the enemy in battle. He was led to this decision by two circumstances, the weakness of the fort at Amherstburg and the success which attended the efforts put forth by his agents and assistants in the mobilizing of a resisting army.

The garrison force under the command of Colonel St. George, when the American army appeared in the district, consisted of two hundred and fifty in total strength, comprised of two hundred men of the Forty First Regiment, and fifty of the Royal Newfoundland Fencibles. Cooped up in the Fort, their situation there could hardly be called one of military strength. "The fort of Amherstburg could not have sustained a siege of any duration," a later member of their regiment wrote.* "Quadrangular in form, four bastions alone flanked a dry ditch, offering little obstacle to a determined enemy. This passed, a single line of picketing, perforated with loopholes for musketry, and supported by a slight breastwork,

---

*Major John Richardson.

remained to be carried. A prudent commander, would, however, have chosen a less uncertain mode of dislodging the garrison. A few shells, promptly directed, would have answered the purpose, since, with the exception of the magazine, all the buildings within were of wood, and covered with pine shingles of such extreme thinness as would have been found incapable of resisting missiles of far less weight. The disadvantage of awaiting the enemy in this position, Colonel St. George well knew. He consequently preferred giving him battle with the trifling force he had at his disposition. With this view, the garrison received orders to be under arms at a moment's warning, and the approach of the invader was anxiously awaited."

Colonel St. George was not, however, fated to meet General Hull in battle. The loss of his incoming military correspondence at the hands of Lieutenant Frederic Rolette on water, and his outgoing mail by the successful assault upon Major Van Horne's escort, by a band of Indians under Tecumseh and Captain Elliott, caused the American General to pause before taking a decisive step to carry out his pre-arranged plans of attack upon Fort Amherstburg, plans, with no longer any element of secrecy about them, but which now had become common knowledge to Canadian leadership. His communications, both by water and by land were cut off. He must therefore open out these before he could initiate an aggressive campaign. This delay which occurred in carrying out the pre-arranged American plan of campaign was not of General Hull's choosing. It was forced upon him by the successful activities of the Canadian defensive forces under Colonel St. George. Detroit was isolated. The Canadian defensive forces were stationed at the most strategic point available. They stood between his army and his supplies. Eleven boat-loads of his provisions had been captured by that fearless and aggressive French-Canadian, Lieutenant Frederic Rolette. Additional stores at the Miami were remaining there stationary, because they could not bring them forward, lest an alike catastrophe would befall them. Shawnee Indians under Tecumseh, Hurons under Chief Roundhead, and the Pottawatamies guided by equally able Chiefs were stationed to attack moving bodies of their soldiers on the right bank of the river, while the Garrison and Militia protected the left. The Queen Charlotte, with Captain Hall in command, and the Hunter, under Rolette, these two vessels of the Provincial Navy were stationed on the river, and waiting in readiness to aid the transportation of the forces on either side to whatever place most and soonest needed. The American Commander was out-Generalled. Colonel St. George could not make any boast of numbers in comparison with those of the American army, but when he separated General Hull from his supplies, and had stationed his men in positions, which would enable them to maintain this point of advantage for an indefinite period, he saved Fort Amherstburg from assault and his garrison army from defeat. Unfortunately the services of this efficient leader were cut off in the early stages of the Campaign. Wounded five times in leading his Brigade—The Newfoundland Fencibles and Essex and Kent Militia—to successful victory at the battle of Frenchtown, he was incapacitated for further service, and returned to Great Britain for recovery, leaving behind him a reputation renowned for efficient leadership and courageous endurance.

II.

PRIVATE HANCOCK

The Soldier Hero of The Canard River

SOMEWHAT past midway between Sandwich and Amherstburg, a sluggish stream finds its way to, and empties into the Detroit river. There is nothing picturesque about this stream, nothing to attract the admiration of a traveller. Its banks are low where it empties into the Detroit river, and the immediate lands bordering on it, are low-lying, flat and marshy. To a stranger, unacquainted with our history, this river would present no picture, other than an environment well-suited to be the home of bull-frogs, and the nesting-place of red-winged grackles.

Yet the Canard river, for that is its name, so devoid in itself of any attractive feature, is lifted from the category of the commonplace by reason of its historic associations. Somewhere near its banks, in the beginning weeks of the war of the American Invasion, a British soldier consecrated the ground as an historic shrine, to be commemorated forever in the annals of Canadian story. He was not only the first to shed his blood in that war, the pioneer among the fallen, but this, under circumstances which displayed such conspicuous bravery and unyielding devotion to duty, as to give him the right to hold a perpetual place of honour in the scroll of our country's history.

Twenty days after his arrival in Detroit, General Hull moved his army across the river and established himself in Canadian territory, in a settlement estimated at that time to contain about twelve hundred people, including the village of Sandwich. Many of the men of the settlement eligible for

215

service had moved to Amherstburg to join the ranks of the Canadian militia, being assembled there. Once established, he issued a proclamation addressed to the 'inhabitants of Canada,' in which after recounting the so-called vices of Great Britain, and the virtues of the Republic, he invited them to become a part of the Union. He promised them, if submissive, the protection of the United States of America, but if they refused this kindly protection, then his army, the vanguard of others to follow, would be used to crush any armed resistance. No quarter would be given those making alliance and fighting with the Indians. To them, there was no alternative, only extermination.

But the display of this great force, and the assurance of General Hull's promises, and threats, did not obtain the response from the inhabitants of the Western district it was confidently expected it should. The disbanded British soldier and the United Empire Loyalist, with the memory of confiscated properties, was not making a voluntary concession of any kind, much less hastening to become a subject of Republican America. For them, the threat of extermination carried with it no fear. They would ally themselves with Indians or any others who would join with them to save their country, their properties, and their homes. Few in number, they were yet strong in spirit, and battles are not always won by merely brute strength.

This message of the United States, having been delivered to the people, General Hull detached foraging parties from his army units, and sent them forth to visit the French Canadians in their immediate neighbourhood, and the British and German settlers on the banks of the Thames, requesting them to pay tribute to the invaders in the form of cattle, sheep, corn, swine and wheat. In this way was solved the problem of the American commissary to obtain needed provisions for their army.

In the meantime, that energetic and spirited commander of Fort Amherstburg, Colonel St. George, was not without resource. To prevent these foraging parties from making sallies on the Lake Erie settlement, and to hinder the march of Hull's army, if it should purpose an attack on their garrison post, he had the bridge over the Canard partially destroyed. The river, where it is crossed by the highway, is three or four rods wide, and of such a depth as to make it impassable for even cavalry, except by the medium of bridge or boat. A gunboat, the Queen Charlotte, was stationed at the mouth of the river to harass any who might attempt to repair the bridge. In addition, and in answer to General Hull's threat of extermination, in respect to Indian alliances, detachments, now of Indians, and again of whitemen, were set in ambuscade among the tall grass and rushes of its banks, to keep watch at intervals over this strategic spot. Daily skirmishes took place from the fifteenth to the twentieth of the month between these ambuscades and detachments from the American army sent forward to reconnoitre and repair the bridge. On one of these occasions, and in accord with the irony of fate, it fell to the lot of a band of twenty Menomini warriors, American Indians from Green Bay, Michigan, to man this outpost or picket. Two hundred of Hull's army approached. The grass and rushes could hide the Indians from observation, but not the Americans. When the detachment reached the place of ambuscade, they were met by a fusilade which compelled them to fall back hastily, completely routed and defeated, an impressive notice to General Hull concerning Indian alliances.

But it is the mission of this chapter, not to record Indian successes, however great a factor they were in the defence of the Canadian Western Frontier, but the answer of the British soldier to the over-running of the country by an American army with a view to its subjugation and conquest. "The standard of Union now waves over the territory of Canada," proclaimed General Hull when he had penetrated as far east as the bank of the Detroit river. The hero of the Canard answered, "If your army shall pass over this highway, it shall be over the dead body of the British soldier posted here to defend it."

On one occasion a certain inferior officer had the Canard picket in charge. Two of his men, Private Hancock and Private Dean, he had placed in forward positions as sentinel scouts or guards. When he deemed it expedient that his detachment should return to Fort Malden, and orders were given to that effect, he committed the careless blunder of neglecting to notify these men of his purpose, who, therefore, continued their sentry duties, though quite unaware that they were left there alone. In the meantime, a detachment of Americans appeared on the scene, initialled an attack, which the two sentinels withstood, nothing daunted. Doubtless they were in expectation that their comrades would be on the spot instantly to assist them, but the rushes and the tall grasses of the neighbourhood brought them no aid.

Two against two hundred! What a disparity in numbers! There could be only one result under such circumstances. Dean was the first to fall. When wounded, he ceased firing and surrendered himself to the enemy. Not so, his companion. With the unyielding spirit of his race, he put up the flag of "NO SURRENDER." Desperately wounded in two places, his arm broken, and on his knees from lack of strength to stand, he resisted with bayonet "the advance of a body of men, who had not magnanimity enough to spare the life of so valiant and resolute, yet so helpless a foe." He asked not for quarters and he received none. When his comrades again visited the spot, they found his mutilated body pierced with many wounds from the weapons of many American soldiers.

This display of heroism has peculiar interest to Canadians, since Hancock stands out in the calendar of time as the first to make sacrifice of his life in the first war we were forced to undertake for the defence of our country. Such an unyielding spirit of devotion to duty, so heroic an example of the qualities of the race from which he sprang, has a right to be perpetuated in the song and story of our people.

It was not the quantity of men that saved Canada from being wrested from British connection on the occasion of that invasion. It was the spirit of her people. Of that spirit, the soldier Hancock stands out an illustrious example. So long as the country can produce men, who, as he, would sacrifice life itself, before being guilty of dereliction of duty, so long will the country be safe and prosperous.

Geography, history, and commerce have joined with the sentiments of men, dividing the world up into national groups, each with its own country and its own political institutions. No better method has been devised for obtaining allegiance to law and order. Since fate has decreed for Canada a destiny of its own, Canadians have reason to cherish with love and enthusiasm the memory of one who gave his life in order that his fellow-countrymen might not be compelled to accept a form of government and a system of rule, which was not in accord with their own sentiments and choosing.

Concerning Private Dean, his companion, the first among the prisoners taken, as Private Hancock was the first among the killed, his liberation took place shortly afterwards under circumstances as gratifying to himself as they were pleasing to his compatriots in war. As soon as the British troops took possession of Fort Detroit, "the first act of General Brock," wrote the historian, Richardson, concerning this incident, "was to enter and liberate, in person, the gallant Dean who had been taken prisoner at the Canard and who then lay confined in the guard room. Shaking him by the hand in presence of his comrades—while his voice betrayed strong emotion, he warmly approved his conduct, and declared that he was indeed an honour to the profession of a soldier."

An official report, concerning the event reads as follows: "The Commander of the Forces takes great pleasure in announcing to the troops, that the enemy under Brigadier General Hull have been repulsed in three attacks made on the 18th, 19th and 20th of last month, upon part of the Garrison of Amherstburg, on the river Canard, in the neighbourhood of that place; in which attacks His Majesty's 41st Regiment have particularly distinguished themselves. In justice to that corps, his Excellency wishes particularly to call the attention of the Troops to the heroism, and self-devotion displayed by two privates, who being left as sentinels when the party to which they belonged had retired, contrived to maintain their station against the whole of the enemy's force, until they both fell, when one of them, whose arm had been broken, again raising himself, opposed with his bayonet those advancing against him, until overpowered by numbers."

—Extract from General Order, Quebec, Aug. 6th, 1812.

III.

GENERAL SIR ISAAC BROCK

His Defeat of America's Initial Plans for the Invasion and Conquest of the Western Frontier

UNPREJUDICED historians are one in their belief, that nothing occurred of so favourable a consequence to Canadians in the events connected with the War of the American Invasion, as the surrender of Detroit on that memorable date, August 16th, 1812. This was not because the Americans lost ultimately any territory thereby, or perhaps even military prestige. It gave, however, to our invaded country a much needed respite until assistance came to us both from the Indian tribes of the Lake Region and the Canadian militia. The significance of this cannot be ignored. It meant the defeat of the enemy's initial plans for the invasion and conquest of the Western Frontier, a failure which secured for Canadians the independence of their country and its continuance as a self-governing part of the British Empire. The honour of having occasioned this surrender, and the consequent defeat of the American plans must be accredited to General, Sir Isaac Brock, the President and Administrator of the Government of Upper Canada, and the Commander that day of the Canadian troops.

Major-General, Sir Isaac Brock, was born at Guernsey, 1769, the eighth son of John Brock of that island. At fifteen years of age he joined the British army, and from the first gave promise of future achievement. He came to Canada in 1802, holding the rank of Lieutenant-Colonel, and to his credit service at the Barbadoes, Holland and the Baltic, which earned

221

for him both distinction and promotion. In Canada, he was appointed to the Command of the troops in 1806, and raised to the rank of Major-General in 1811. That same year he was appointed to the office of President and Administrator of Upper Canada, a position which he held when the Americans declared war against us in 1812. Although his service in the Western Frontier was only a matter of days or weeks, he nevertheless left for himself, as a result of his splendid achievement, in occasioning the surrender of Fort Detroit, a place of undying fame in the history of this district, this province and of Canada.

Although war had been declared more than six weeks previously, it was towards the end of the first week of August before General Brock was ready to leave Niagara for Detroit. He arrived at Amherstburg on the morning of the thirteenth, and with him a force of two hundred and sixty men of the Canadian militia, chiefly volunteers from Toronto. The next two days were spent in marshalling the forces at his disposal, and the assembling of them in the neighbourhood of Sandwich. On the same day, and on the shore three miles east of Sandwich opposite the fort at Detroit, there was completed the establishment of a battery, the preparation for which had been begun, after the evacuation of the Canadian shore by General Hull, a week previously. These preparations finished, Brock demanded the capitulation of the Fort and the surrender of the army from General Hull. This being refused, he crossed the river the next morning with his army, prepared to give battle to the American forces assembled within and in the neighbourhood of Fort Detroit.

The crossing of the river by the Canadian army on the morning of the sixteenth has been succinctly described to us by John Richardson, the Amherstburg schoolboy of fifteen, who was then a part of the Canadian defensive forces gathered together so hurriedly for the salvation of the country from subjugation and invasion:

"The batteries which had kept up an irregular fire during the night, renewed it at the first dawn with unabated spirit, and the requisite boats having been provided, the crossing was effected without opposition, under cover of the guns of the Queen Charlotte and General Hunter, which lay anchored about half a mile above Sandwich. A soft August sun was just rising as we gained the centre of the river, and the view, at the moment, was certainly very animated and exciting, for, amid the little squadron of boats and scows, conveying the troops and artillery, were mixed numerous canoes filled with Indian warriors, decorated in their half-nakedness for the occasion, and uttering yells of mingled defiance of their foes and encouragement of the soldiery. Above us again were to be seen and heard the flashes and thunder of the artillery from our batteries, which, as on the preceding day, was but feebly replied to by the enemy, while the gay flags of the Queen Charlotte, drooping in the breezeless, yet not oppressive air, and playing on the calm surface of the river, seemed to give earnest of success, and inspirited every bosom."

The army was safely landed without opposition, and arranged in a position favourable to assault, with General Brock in advance reconnoitering the fort, when a white flag was seen advancing, the purport of which it was soon ascertained was an offer on the part of the Americans to capitulate and to surrender the fort. This turn of events was assuredly as gratifying as it was unexpected.

General Hull has been greatly blamed because he

permitted the surrender to take place, but, perhaps, unjustly. A careful survey of the whole situation by an unprejudiced mind, might lead to the conclusion that he acted the part of wisdom in following the course which he did. It was one of those military events, the explanation of whose results is to be found in psychological as well as in military reasons. In warfare, as in all the great events of life, the unexpected often happens. In the Great World War, Germany's magnificent military organization, so efficient and resourceful, would seem to prognosticate with certitude, its ultimate success. But this efficient and powerful organization did not win out in the end.

In the war of the American Invasion, the preparedness and strength of her military army in the Western Frontier, was pregnant with promise for the immediate capitulation of Upper Canada. Their army of twenty three hundred men, which they had forwarded into the Detroit district, to await in readiness the declaration of war, was three times stronger than anything which Canada could immediately foregather to meet them, Fort Detroit had been prepared and put in good order the year before. It was amply supplied with guns, cannon and ammunition. It was garrisoned by a strong force. The army took its place under the shelter of this fort. All this took place before the little garrison at Amherstburg had any assured intelligence that war with the United States was to take place, much less to afford them adequate preparation for any kind of defence. Yet, notwithstanding all these favouring circumstances, General Hull, the prepared aggressor, surrendered the fort, his army and the surrounding territory into the hands of Sir Isaac Brock, the invaded country's General, an event as unexpected to the American nation as it was unwelcome.

MONUMENT TO BROCK

BROCK'S MONUMENT

On the 14th of March, 1815, the Legislature granted by Act one thousand pounds "for the constructing of a monument to the memory of Major-General Sir Isaac Brock" and on January 30th, 1826, a further sum of six hundred pounds was granted to complete it.

On the 13th of October, 1824 the remains of Major-General Brock and his gallant aid-de-camp, Colonel John Macdonell, were taken from a bastion of Fort George, where they were first interred, and placed in a vault beneath the monument.

On Good Friday, April 17th, 1840, this monument was irreparably wrecked by a rebel of 1837 named Lett, who had managed to explode within it a quantity of gunpowder. A monster meeting was called for the next 30th of July at Queenston Heights, for the purpose of adopting measures for the erection of another monument. The meeting was a great success; a building committee was formed, and arrangements were made for soliciting subscriptions from the people of Canada. In 1853 the new monument was begun. The corner-stone was laid on October the 13th, 1853, by Lieut.-Colonel Donald Macdonell, Deputy-Adjutant General of Militia, and brother of the gallant officer who had shared in the glory and met the fate of General Brock, forty one years before. On this day the remains of the two brave men were deposited in two massive stone sarcophagi in the vault prepared for them in the new structure.

Major-General Sir Isaac Brock, the eighth son of John Brock, was born in the island of Guernsey on October 6th, 1769. In his fifteenth year, March 2nd, 1785, he purchased an ensigncy in the 8th King's Regiment, in 1790 was promoted to a Lieutenancy and at the close of the year was Captain. Soon after he exchanged into the 49th, then at Barbadoes, where he remained till 1793, when he returned to England on sick leave. He purchased his Majority on June 26th, 1793, and on October 27th, 1797, purchased the Lieutenant-Colonelcy. He served in Holland in 1799, and in the Baltic in 1801, and came to Canada in 1802. He was made Colonel in 1805, and, after a visit to England, succeeded to the command of the troops in Canada in September, 1806. On June 4th, 1811, he was appointed a Major-General on the staff in North America, and on September 30th, was appointed President and Administrator of the Government of Upper Canada, an office he held at the time of his melancholy death, at Queenston Heights, Niagara river, while resisting successfully the invasion of Canada by the Americans, October 13th, 1812.

There were four principal reasons which led General Hull to do this, two of them military, but two of them psychological.

The most potent psychological reason was the aversion of General Hull to the Indian and his method of warfare. He could not but know, that notwithstanding his gestures of power, and the efforts of his people to induce them to stand aside neutral, these were joining the Canadian forces with rapidly increasing numbers every day. If not already equal to his own army in numbers, he adjudged that they soon would be. In the immediate neighbourhood of Detroit alone, although not then as numerous as when France and Britain possessed that post, the four tribes of Hurons, Ottawas, Ojibways, and Pottawatamies, could muster a considerable body of men. There were besides the warriors of the Shawnee chief, Tecumseh, from the Wabash, who had readily come forward, when they found that a new opportunity was to be given them to attack the nation which had wrested from them their hunting grounds north of the Ohio. Menominis from Green Bay, Michigan, had already tried their mettle against a detachment of his army at the Canard and worsted them. Then, there were those who were on their way there, coming by canoe from the far north, from the shores of Lake Superior, the islands of Lake Huron, and all that vast area extending from the north shore of Lake Huron to the Hudson Bay. But nearer to his own home, the tribes driven north by the armies of General Wayne, the Miamis, these were not forgetful of the past, and under their own chiefs were ready to stand side by side with the prophet's brother to support the Canadian cause against America. All that these Indians needed was efficient leadership, equipment and provisions and there would be available a sufficient army to maintain a successful defence of the Western Frontier of Upper Canada.

But there was about these Indians that which General Hull disliked more than their numbers—their method of warfare. The Indian knew nothing of the canons of so-called civilized warfare. According to this code of warfare to shoot a man with a gun in his hand is right procedure, but to shoot him unarmed or in a state of helplessness, this is cruelty and murder. The Indian was not so tutored. With him, an enemy was an enemy, to be despatched whenever and wherever possible, the more defenceless the better. His method was thorough, relentless, and cruel. It acknowledged no difference as between men, women and helpless children. His glory was to get scalps without any embrasures to his own skin. Kill and scalp and run—that was scientific warfare with the Indian. Moreover, the savage instinct once let loose, the lust for blood once inspired, the mob method once encouraged, and there was no White leadership strong enough to prevent them from carrying out their savage instincts. General Hull was quite aware of all this, and lest there was any danger of his overlooking these traits of Indian character, he was reminded of them by General Brock.

But in addition to the psychological, there were military reasons why General Hull became predisposed to accept capitulation instead of battle. There was first, the effectiveness of the battery established on the Canadian side, opposite Detroit. This, under the command of Captain George B. Hall, Commodore of the naval fleet, and manned by the marines, kept up a fire throughout the whole of the previous night, though at irregular intervals. At daybreak, it became a spirited attack. As an instance of the effectiveness of the work which was done, one ball of eighteen pounds entered into the mess room, and killed four officers, of whom Lieutenant Hanks, who had surrendered Fort Mackinac, was one of them. This battery could be easily turned on the army when it came out to counter the attack on Brock against them, and be counted upon doing effective work there also. This was an added reason why he should consider carefully the offer of capitulation.

But the more potent military influence, affecting the purpose of General Hull, was the generalship of Brock. The intrepid courage, the ready energy, the promptitude of execution with which he set out to attack the American army and fortress could not do anything other than stir misgivings in the mind of General Hull and produce doubts as to the successful issue of his own undertakings. General Brock moved as one who knew what he wished to obtain and was confident he could secure it. Such generalship could be counted on getting the best possible results from his available resources. There was that about the whole movements staged by General Brock which convinced him that, should a battle or an assault be staged, there would be no witholding of the hand of conflict until the last available resources of the Canadian army were completely exhausted. Discretion, he knew to be the better part of valor, and he played this better part.

"In the capture of Detroit, General Brock has been termed the saviour of Canada, and most deservedly so. Had he not struck the blow he did, and at the time he did, at the American power in the West, Upper Canada—nay both of the Canadas must have been yielded to the triumphant arms of the United States." . . . "He well knew that, on the destruction or the discomfiture of the North Western Army, depended the safety of the province committed to his charge, and the enterprise, which he himself has termed hazardous, was perilled only after profound reflection and conviction. He justly entertained the belief that while, on the one hand, the slightest delay and incertitude of action, would be fatal to the interests of Great Britain, inasmuch as it must have tended to discourage, not only the inhabitants of the province, but our Indian allies, there was, on the other, every probability, that an immediate and vigorous attack on an enemy who had already lost so much time in inactivity, and who had abandoned so many advantages, would be crowned with success. It was a bold—an almost dangerous measure; but the danger of the country was greater, and he resolved to try the issue. He succeeded; from that hour Canada was saved."*

The service which General Brock thus rendered, in the Western Frontier was enhanced in the esteem of his fellow-countrymen by reason of the fate which befel him at the battle of Queenston Heights. What he did there was added testimony, if such were needed, of what he would have done at Fort Detroit on the sixteenth, if a combat had been demanded of him. With such a leader, not a man would be found, either among the British Regulars, the Canadian militia, or the Indians, who would not have followed his example with their utmost endeavour. The General who ordered the capitulation, and the soldier who prepared and wrote out its terms, both of them a few weeks later faced death and accepted it as an alternative, rather than permit this country to be brought into subjugation by the superior power of an hostile invading army.

---

*War of 1812,—by Major John Richardson.

IV.

ROBERT DICKSON
Explorer, Indian Trader

His Service in Behalf of the Canadian Defence of the West
1812-13

ROBERT DICKSON was born in Dumfrieshire, Scotland, in 1768. He came out to Canada when a very young man, and devoted his life to the vocation of explorer and trader among the Indians until his death at Drummond Island, June 20th, 1823. He was one of the first whitemen to ascend the Missouri river to its source, and was intimately acquainted with every Indian tribe inhabiting the great North West. He acquired a remarkable influence over the Menominis, the Winnebagoes and the Sioux, being in many instances the first whiteman to trade among these hitherto, unknown tribes of the far west. He was a staunch friend of the Indians, and an indefatigable trader among them during all those years in which they and the United States were at war for possession of the great North West territory. Though his sympathies were strongly on the side of the Indians, yet he used his influence on all occasions to induce them to seek peace with the United States and cease from committing depredations on their frontier settlements, giving them occasion to put their policy of extermination into effect. His sympathies in Indian favour were enhanced by reason of his domestic relationships. He was married to an Indian woman. His children—and he was a man of warm, enduring, filial affections—all contained Indian blood equal to that of his own race. In his business relations he became widely known among the most respectable families of the province, and because of his wide range of knowledge concerning the Indian tribes, and his influence over them his aid was sought on more than one occasion by the Government to effect important negotiations with them.

The winter preceding the United States invasion of Canada, he had been sent on a mission of aid to the Mississippi and Missouri Indians, who were reported in a condition of great distress from lack of food and ammunition. The preceding summer was marked by an unusually long drawn out season of drouth, which had destroyed their harvests and drove the animals of chase farther north out of their district in search for food. This loss of harvests and migration of the deer brought many of the tribes to a state of destitution and without food to meet the needs of the approaching winter. In sending Mr. Dickson to assist them in their need, this is not to be construed as anything other than a purely business arrangement on the part of the fur-traders, the prosperity of their business being dependent upon the prosperity and activity of the Indians. But while on the part of the Companies, it was only a matter of business, yet in respect to the Indians, it was a matter of humane generosity, their appreciation of it being expressed in sentiments of loyalty and good-will towards Mr. Dickson.

A person enjoying the good-will of the Indians and with business capacity to oversee the distribution of provisions for so many tribes, and with that personal knowledge of their number as would enable him to safeguard the repayment to the Company subsequently with furs, such a person would be an invaluable aid to the Canadian Government to obtain Indian alliance to defend their country in the event of an outbreak of war. It is not to be wondered at, then, that his exertions in aid of Canadian defence were solicited as early as the month of February preceding the opening out of hostilities.

In a confidential letter communicated to him from Captain Glegg, enquiry was made as to the disposition of the Indians towards the Canadians; and whether in the event of war with the United States, which was now an assured probability, what tribes of them would be induced to come to the aid of Canadian defence. This communication arrived at a time when Mr. Dickson was making some discoveries on his own account. In respect to every tribe which he visited, he found that American agents had either preceded or followed him, distributing gifts with lavish generosity. As this was a new procedure on their part, he rightly concluded that it was an evidence that the United States authorities meditated aggressive hostilities against Canada and were taking these advance measures to assure victory. But the personal influence of Mr. Dickson over these far west tribes was too great to be superseded by a lavish distribution of American gifts. To be prepared for the expected emergency, he proceeded in the month of May to recruit a certain number from among the Sioux whom he had directed to assemble themselves at Fort St. Joseph, an undertaking which succeeded in winning over the Ottawas of that neighbourhood on the Canadian side. At the same time, he sent a band of Menominis from Le Baye to strengthen the garrison force at Amherstburg.

More credit is due Mr. Dickson for the services which he rendered in securing Indian aid than has yet been accorded him. We must not lose sight of the strenuous efforts which had been put forth by the United States to gain the alliance of the Indians as active participants in the war on their side, or failing in this to have them stand aside, neutral. Nor are we to come to the hasty conclusion that no success attended these efforts. It is true that the battle of Tippecanoe had alienated Tecumseh, and with him a small band of the Shawnees tribe over whom he was chief, but, on the other hand, when war was declared, the Hurons had refused to ally themselves on the side of the Canadians, the Ottawas were standing aloof at Mackinac and Fort St. Joseph, and the Five Nations, in the neighbourhood of Brantford, had openly declared themselves on the side of the Americans. It was Mr. Dickson who was the first to make a breach in these relationships, and the most active and successful man throughout the campaign in summoning the Indians to Canadian aid. "Among the individuals who exerted themselves on the occasion with so much spirit and ability, the first place is generally allowed to Mr. Robert Dickson."*

The services of Mr. Dickson were first directed, as has been mentioned, to the strengthening of Fort St. Joseph. This Fort was built and occupied by the British in 1796, when compelled to evacuate Fort Mackinac under the terms of the Jay Treaty of 1794. It was situated on one of the Manitoulin Island group at the north-western corner of Lake Huron, with a view to the protection of navigation on the St. Marys river. At the outbreak of hostilities it was manned by a garrison of fifty regular soldiers with Captain Roberts as Commander. To this post he brought the Sioux which he had mustered from the far west, and they were there in readiness for active warfare when official notice of the opening of hostilities

---

*"A sketch of the British Fur-trade in North America"—by the Earl of Selkirk, London, 1816.

arrived at the Fort. In the accompanying message from General Brock, Captain Roberts was urged to take such measures as his own wisdom would dictate, whether of offence or defence, but with promptitude, as the enemy had already mobilized their army on the Detroit river, in readiness for the invasion of Upper Canada, west.

Captain Roberts, on receipt of these instructions, proposed an immediate attack upon Fort Mackinac as the best possible means of defence. During the French regime, this fortress was established on the mainland, but when the British took over the country, they purchased an island off the mainland of Northern Michigan from the Ojibway Indians for £5000, and removed the fort to that location in 1780. This was one of the fur-posts, and next to Detroit, the most important of them ceded to the Americans and evacuated under the terms agreed upon in the Jay Treaty. The American garrison, when General Hull began his invasion of Upper Canada, consisted of sixty soldiers of their regular army with Lieut. Hanks in command.

Several considerations influenced the mind of Captain Roberts to take these offensive measures against the American fortress, now that hostilities had been begun with the Americans as the aggressors. Chief among these, inducing him to act with promptitude in the matter, was the weakness of his own fort on St. Joseph's island and the receipt of information that reinforcements were being forwarded to strengthen the American position. Successful action was needed also, in order to win over the Ottawas to the Canadian side. In addition, the Indians which Mr. Dickson had assembled, must be given an immediate opportunity to meet the enemy, otherwise, in accord with the well known traits of their character, they would disperse, and perhaps abandoning their Canadian alliance, join themselves with the Americans. Immediate action therefore became an imperative duty. A fleet of canoes and sailboats were put into requisition, and with his fifty regular soldiers, one hundred and fifty fur-trading men, and Robert Dickson's Indians, they started out for Mackinac on the morning of the 16th of July, and with them guns for assault purposes should the American garrison refuse to surrender on demand. Twenty four hours after, the fort was surrounded, and a request to surrender sent to Lieutenant Hanks by Captain Roberts. In order to save a fruitless resistance which might mean the massacre of all his force, Lieutenant Hanks without further parley, wisely acceded. Generous terms of capitulation were granted him, and at twelve o'clock, noon, on the stroke of the hour, the American flag was hauled down, and the British hoisted in its place, and in that position of triumph, it remained until the end of the war. Several serious attempts were made at later dates to recapture it, but these ended in every instance in failure, evidencing the ability of the Western Frontier to adequately defend itself, if given the required leadership.

Immediately following this initial success, the Indians brought there by Mr. Dickson, were returned to their own country, though in the meantime an appeal came from General Brock for aid to be sent to Amherstburg. Hence only the Ottawas and the Ojibways remained, and these were as yet, only half enthusiastic in their loyalty.*

---

*"The General must have imagined that all the Indians brought here by Mr. Dickson, as well as those that had been collected for the expedition against this place, were still here. The dreadful consumption of provisions caused by these people, who flocked in from all quarters with their wives and children, obliged me to send them off as fast as possible, and the distant Indians were no sooner served with presents than they were warned to return to their country."
               Letter from Captain Roberts, August 16th, 1812.

The task of recuiting these was assigned to John B. Askin, an Indian Agent and storekeeper at St. Joseph Island. Mr. Askin was the son of the well known John Askin, the pioneer of that name in the Detroit river district. Like Mr. Dickson, Mr. John B. Askin was married to an Indian woman, and wielded a great influence over the Ottawas and Ojibways of that region. He succeeded in enlisting a band of two hundred of these under Big Gun, Little Knife and Assiginack as their Chiefs, and started south with them in canoes, accompanied by about seventy voyageurs, to support the Canadian cause at Amherstburg.*

Between Fort St. Joseph and Amherstburg is a body of water of long distances. To travel this chain of waterways in time of peace would be a hazardous undertaking, but in time of war, with the menace of unfriendly Indians all along their route, the undertaking was doubly hazardous. Impelled by the need of Amherstburg, and with his corps of Indians and voyageurs incited to new ventures by the surrender of Fort Mackinac, Mr. Askin lost no time in starting southward. Something of the celerity with which these men acted and the risks which they ran, can be gauged from the time which they spent in making this journey. Favored with suitable weather, when they came to Saginaw Bay, they struck for the open sea across the bay, a course which led them for many miles out of sight of land. Daring this, and many hazards, they succeeded in making the journey in six days, a feat unparalleled in the history, hitherto, of canoe navigation.

---

*"A young Ottawa, one of the crew, was the bearer of Wampum from a chief called the Wing, in the river St. Clair, to the Ottawas here, telling them they had done wrong in assisting their English Father, that the Americans were as numerous as the sand, and would exterminate them."
Letter from Captain Roberts, Aug. 16th, 1812.

In the meantime, Mr. Dickson moved out to the more remote regions, and gathering many brought them to Detroit, and there succeeded in establishing an Indian army, which was said to include, on one occasion, a force of three thousand warriors. How invaluable the service which he rendered can be realized only when we consider how indispensable the Indian was to a successful defence of western Upper Canada. The small handful of British regulars detailed for defence duty, could be no match for the Americans' thousands, even when reinforced by detachments of the militia from the sourrounding settlements. In respect to zeal, energy, and devotion to duty, the Forty-First regiment leaves nothing to be taken for granted, but how impossible for a force of a few hundred no matter how spirited and efficient, to meet and cope successfully against an army fully equipped with efficient munitions of warfare, which had as many thousands as they had hundreds. The Canadian force of British regulars and militia had to be augmented, if any resistance against the invading hosts of the United States should be attempted. This supplement of men could be obtained only from among the employees of the trading companies and the Indians. Had the Americans won the Indians over on their side, the conquest of Upper Canada by them would have been a matter of words, not deeds. To discover then the source of responsibility for the Americans' failure to meet with their expectations on the Western Frontier, the historian will have to go further back than the Indian. He will find that the inspiring medium for the strong arm of resistance which they met on the front line of Canadian defence was provided by such men as Robert Dickson, John B. Askin, Angus McIntosh, Lewis Crawford and the many others of their class associated with them. If the

Indian was an essential factor in every engagement in which
the invading armies of America were successfully resisted, and
we have no occasion for denying his presence in adequate
numbers, or belittling the importance of the service which he
rendered, then to these men belong the credit, for they were
the ones in the main, responsible for his indispensable presence.

There was also another field in which the name of Robert
Dickson has come down to us in record as having given great
and creditable service. This indispensable element, the Indian,
required close and careful handling. He had to be kept above
all things else from excesses. He had to be tutored in Canadian
methods of warfare. The Indian aim in war was not to overcome the enemy, but, to kill. It was not battles won that
counted with them, but, scalps. Offers of surrender, conditions
of helplessness, appeals for mercy, these brought no response
if their savage instinct was not kept in restraint through the
influence of whitemen whom they were accustomed to obey.
There were many white leaders who could hold them in hand,
and half-breeds also, but none apparently could surpass Mr.
Dickson in influence over them. "After the arrival of Mr. R.
Dickson" reported General Procter on one occasion concerning
him, "his Indians were restrainable and tractable to a degree
that I could not have conceived possible." Others associated
with him in the work have written concerning his influence
with similarly approving commendation. When victorious and
given access to the enemy's supplies, in which rum was oftentimes a commodity, without some restraining influence, there
would be exhibited by them the savagery of the mob in its
worst and most baneful extreme. More than one instance
occurred in which they did get out of hand, exemplifying the
reprobate savagery of their nature, to the utter loathing and

reprehension of those compelled, by the force of circumstances, to be associated with them in warfare. Reprehensible as these instances were, their number would have been much greater had it not been for the beneficent service rendered by Robert Dickson and others associated with him in keeping them tractable and under restraint.

The question has been asked, why should Indians have been employed at all, if the keeping them in restraint proved so difficult a problem? This is a question that is on a par with another one, Why do nations go to war at all? To the former of these questions, the historian, Major John Richardson has given answer:

"The natives must have been our friends or our foes; had we not employed them, the Americans would; and although humanity may deplore the necessity imposed by the very invader himself, of counting them among our allies, and combating at their side,—the law of self-preservation was our guide, and scrupulous indeed must be the power that would have hesitated at such a moment in its choice. The act of aggression was not ours—we declared no war against America —we levied no armies to invade her soil, and carry desolation wherever they came,—but we availed ourselves of that right, common to every weak power—the right of repelling acts of aggression by every means within our reach."

Since circumstances decreed that Canadians should be made dependent upon Indian aid in the defence of their country, it is a matter of self-congratulation that both officers and civilians united to keep the Indian tractable and restrained. One of the most valued of Indian exhibits in possession of our historical society is a sword which was presented to Little Knife, an Ottawa chief, by the Commandant at Mackinac for his kindness to an American officer whom he discovered wounded and saved him from the scalping knife of his followers.

With a knowledge of Indian character, unsurpassed by any others, and with an influence over them which enabled him to keep them in restraint, Mr. Dickson served humanity in equal measure with which he served his country; in the latter, by summoning his Indian friends to the aid of Canada; and in the former, by the integrity of his own personal character, which forbade him to allow any Indian under his charge to strike an unnecessary blow against the wounded, defeated, or helpless, from among the ranks of their opponents in battle.

V.

MATTHEW ELLIOTT

THE most prominent personage in the Canadian Detroit river district, when the war broke out in 1812, was without doubt, Colonel Matthew Elliott of Amherstburg. He was at that time one of the oldest living residents in the district, and held the prestige which comes to a man of long-continued prominence before the public. He also held the prestige which comes from a successful business career.

On the mainland facing the south end of Bois Blanc island at the mouth of the Detroit river, he had selected a homestead of two thousand acres, one of the first land transactions to take place in Malden. There he produced on this side of the Atlantic, the imposing grandeur of an Old Country estate. It was all cleared, and operated profitably, by a retinue of servants and slaves, one of the few creations of its kind in Canada. This vocation, had he no other, would have given him prominence above the average in the district.

But, in addition to his private enterprise, he held at the outbreak of the war, two public offices which brought him prominence and prestige, the County Lieutenant of Essex, and Deputy Superintendent of Indian Affairs. He came to the Detroit river frontier in company with Alexander McKee and others at the commencement of hostilities in connection with the war of the American Revolution. Of Irish descent, he was born in Maryland in 1739, and became a trader among the Indians at a very early age. He was a resident of Fort Pitt when the war broke out, and had aligned himself on the side of the Constitutionalists in support of continued British connection. Threatened with imprisonment he fled to Canada, in company with six others, and took up residence at Detroit. There he was given a position in the Indian Department, by Lieutenant-Governor Hamilton, first as an Interpreter, but later raised to the rank of Captain. In 1780, he accompained Captain Henry Bird of the 8th Regiment, in his raid into Kentucky; and afterwards commanded the western Indians in the actions of the Blue Licks and Sandusky, in which the frontiersmen of Kentucky and Pennsylvania, were defeated with severe loss. His services were recognized by his appointment to the post of Assistant Agent for the western Indians in 1790, on the occasion of the death of Alexander McKee, his former fellow-citizen and associate at Pittsburg, an office which gave to him commanding influence over the Indians both in times of peace and war.

He had also been elected a member of Parliament for Essex county in 1801, an honor which his constituents

QUEENSTON, 1812

On the Canadian side of the Niagara river, just where its foaming and turbulent waters issue from the narrow rocky gorge, stands the straggling village of Queenston. The place at the present time is of very little importance except as a terminal port for a magnificent fleet of pleasure vessels that carry tourists and excursion parties to visit the Falls, five or six miles up the river. But as the scene of one of the proudest victories of Canadian and British arms during the War of 1812 Queenston won a fame that is world-wide.

GENERAL BROCK—"In person he was tall, stout and inclining to corpulency: he was of fair and florid complexion, had a large forehead, full face, but not prominent features, rather small, greyish-blue eyes, with a very slight cast in one of them,—small mouth with pleasing smile and good teeth. In manner he was exceedingly affable and gentlemanly, of a cheerful and social habit, partial to dancing, and although never married, extremely devoted to female society. Of the chivalry of his nature and the soundness of his judgment all comment thereon in view of what he did is a matter of supererogation."

continued to confer upon him during the remaining years of his life. Any one of these positions would have given him a place of prominence in the community, but in the enjoyment of all three, he was raised to a place of distinction, second to no other in the Malden district.

An incident occurred, when a renewed war between Great Britain and the United States was deemed imminent, which shows how great was the influence he was reputed to have among the Indians, especially over the Hurons dwelling at that time on Bois Blanc island. In consequence of a bitter quarrel with Captain Hector McLean of the Royal Canadian Volunteers, the commandant at Amherstburg, 1798, he was summarily dismissed from his position as Deputy Superintendent of Indian affairs. A few years later, Lieutenant-Governor, Sir Francis Gore, urgently requested that he should be reinstated, affirming that he above all western men, had the necessary influence to induce the Indians to align themselves on the Canadian side, when the expected war should break out. Agreeable to the Governor's solicitations, he was re-appointed to the post, a wise movement in view of the active efforts put forth by the United States to effect a breach between the continued friendship of the Indians with the Canadians, especially those dwelling in the newly-ceded American territory.

There cannot be any doubt that the problem of the Indians was one of the most irritating with which the United States had to deal in this period of its history. It was also the one of first national importance to Canada. The complaint of the United States nation is summed up in a paragraph of a letter sent by President Washington to his plenipotentiary, John Jay, while negotiations were pending in 1794: "We have a thousand corroborating circumstances, and indeed almost as many evidences to prove that they are seducing from our alliances, tribes that hitherto have been kept in peace and friendship with us, at a heavy expense, and who have no cause of complaint, except pretended ones of their own creating; while they keep in a state of irritation the tribes who are hostile to us, and instigating those who know little of us, or we of them, to unite in the war against us; and while it is an undeniable fact that they are furnishing the whole with arms, ammunition, clothing, and even provisions to carry on the war, I might go further and, if they are not much belied, add men also in disguise."

All of this is true, but the men who were behind those movements, were the government officials, neither of Great Britain nor of Canada, as was the belief of President Washington, but individual and private citizens dwelling at that time in the ceded North West Territory. Some of them were natives of the United States as Alexander McKee and Matthew Elliott; some of them were natives of Scotland, as Robert Dickson and Lewis Crawford; and some of them were born on Canadian soil, as John B. Askin and Pothier. These men were shaping the future destiny of Canada, as much, or perhaps to a greater extent, than either the statesmen of Great Britain or the United States. They would not deny that they were out to obtain the friendship of the Indians, and their alliance with them to defend the country in the time of war. They would, however, deny to the United States, the right to any possession or claim upon the Indian, whether dwelling in the ceded or unceded territory of Canada. When Great Britain relinquished her claim upon the Great North West, she did not, and could not, cede over to the United States, possession of the people then domiciled in that territory, including the Indians. The claim of the Indians to be considered a free and independent nation, that claim these men supported both in their speech, and actions. They wanted this claim respected by the authorities both of Great Britain and the United States. They based their demand on the inherent right of the Indian to his life and to his liberty, and to the possession of his own lands.

The motive which these men had in supporting this claim of the Indians, was, in the first place, the preservation of the Indian fur-trade. When war was declared by the United States, or when it was viewed as imminent, they supported the Indian claim then, in order to strengthen the military forces available to defend Canada from American Invasion. In both instances, self-interest was without doubt, the in-spring impulse leading them to sympathize with and support the Indian in his effort to maintain his integrity and his inde-pendence. For the United States to find fault with the pro-cedure of these men, would appear to-day on a par with the old adage of Satan rebuking sin. Neither side could claim any great generosity of motive in their relationship with the native population of America, doomed to extinction by the ravages of war and the pestilences of small-pox and whiskey.

The Americans in 1811 became very active in their efforts to prevent the Indians from being used as a force against them, if they should carry out their anticipated war against Great Britain. The battle of Tippecanoe was fought, not only to overthrow Tecumseh's plan of an Indian Confederacy, but to compel them to stand aside neutral during the period of their impending conflict with Great Britain. This was followed by a lavish distribution of gifts among the Miamis and other North West tribes. Their agents were sent out among the Chiefs possessing George-the-Third medals, an insignia of their fealty, who were invited to accept in their place the medals of the Eagle Nation.

But over against these, were the activities of Matthew Elliott and others, with the result that when the war broke out, to the consternation, and entirely contrary to the expectations of General Hull, the Indians, almost without exception, aligned themselves on the side against the United States.

The American Indians stood on an entirely different basis to the war to the Canadian Indians. These latter were wards in perpetuity of the Canadian Government, and though for a time they might be temporary residents of the United States, yet their relationship to the annual presents suffered no change. The war was against their interests, therefore, as well as against those of the British-born settlers. The American Indians, on their own account, and in support of their own policy, that the North West territory should be held the inalienable right of all Indians, seized this favourable opportunity of fighting against the United States. The massacre of the garrison soldiers at Fort Dearborn, had no connection whatever with the Canadian campaign, any more than the battle of Tippecanoe the preceding year. It was a work of American Indians, against the Americans, a part of that long-continued and bloody struggle between the whites and the Aborigines to gain possession of the Great Northwest, the most valued of the Indians' hunting grounds. It was with the Canadian Indians that Colonel Matthew Elliott had to do, though from his knowledge of Indians, and the Indian language, he was called upon frequently to be the intermediary between all of the Indians and the officer commanding the entire army.

The activities of Colonel Matthew Elliott in the war, were such as to give added distinction to his name, if such were needed. At his time of life, he was then 73, he might have been forgiven if he retired from all activities on the outbreak of war, and left the fighting of his country's battle to younger men, but so enthusiastic was he in behalf of the cause which he espoused, and so urgent were the solicitations made by the government to have him take a part, that he took up the additional duties which the war forced upon him with a zeal and devotion, which made him an invaluable asset to the Canadian cause. He was present in person in every engagement in which the Indians took part, during the whole of the western frontier campaign, and this in addition to his duties as County Lieutenant and Commanding Officer of the First Essex Militia Regiment. He had charge of the Indians in the manoeuvres preceding the surrender of Detroit, and of the few prisoners taken by the Indians in the preliminaries of that engagement, he had them carefully protected and suffered no hurt to befall anyone of them. For this service he was later presented with a gold medal. Indeed he was one of the most potent agents in the army during the whole of the campaign to prevent Indian depredations and practises of cruel savagery.

Thus it will be seen that as Commanding Officer of the Essex Militia and as Superintendent of Indian Affairs, he wielded an influence second to no other officer, save that of the Commander, General Procter. But seeing that the plans initiated by this officer were dependent for their success upon Indian co-operation and that Colonel Elliott was the one man above all others qualified to obtain that support, it will be seen that the place of strategic importance which he held was very great.

This will be more forcibly realized when we have a true concept of Indian character. Although indispensable to the Canadian cause, the Indian was as undependable as the weather. He was subject to moods and superstitious fears, was easily discouraged and incapable of a service which was not broken up by frequent withdrawals from the arena of conflict, would seldom fight two battles in immediate succession. In imminent danger, he might or he might not rise up to the occasion but there was nothing in his character that would justify a reliance that his services would be in requisition when they were most needed. He was at his best in spasmodic intervals and in surprise attacks. Colonel Elliott was intimately accquainted with all of these traits of his fallible nature and was therefore constantly on the alert to prevent them from producing disaster, on a strategic occasion, to the Canadian cause.

An illustration of the difficulties facing the Canadian forces dependent upon Indian aid, is related to us in connection with the expedition into the interior, which was undertaken by Major Muir for the reduction of Fort Wayne on the Miami river. This was an American supply depot under guard of about two hundred soldiers. Starting out with about three hundred whitemen, regulars and militia, and with them a body of about five hundred Indians, composed of representatives from the Huron, Mackinac and Saginaw tribes, after a difficult march, they arrived within the near neighbourhood of the Fort only to find that they were meeting an American army under General Winchester, twenty five hundred in strength. Following immediately after them another army of three thousand under General Harrison was approaching. Although apprised of their presence, General Winchester, believing that the Canadian corps were three times stronger than they were, delayed giving them battle until he would be reinforced with Harrison's army. In the meantime, the Mackinac and Saginaw Indians spent the time in conjuring practises, seeking a message from the spirit world as to the outcome of their expedition. Revelation, according to their conjurers' dreams, was given them that they would be defeated. The next morning, irrespective of Colonel Elliott's entreaties, or their promise to Major Muir to support him in battle, in order to escape their revealed doom, they dispersed in bands of sixes or sevens, following one another through the woods, homeward bound.

There remained nothing further for Major Muir and the Hurons to do, other than to return to Amherstburg. This they did in so orderly a manner, that neither a gun or a man was lost by reason of the expedition, though on the other hand, nothing was accomplished, save the knowledge obtained of the strength and prepared conditions of the enemy, and the practise in retreating which the occasion supplied. But had a battle ensued, they would have had to meet the enemy with their Indian support reduced by more than one-half of their numbers. The same dependableness or, lack of it, was exhibited on different occasions during the campaign by various other tribes.

There was said to have been gathered on the side of the Canadian cause, at different times, representatives of at least twenty tribes, and numbering all told between three and four thousand warriors. But these were never present on one occasion and to fight. Some would be there to fight, some to look on, some to conjure, but all of them,—men, women and children—to be fed and kept friendly with rum, a herculean task under the supervision of Colonel Matthew Elliott. Thus, he acted as intermediary, sometimes successful, sometimes unsuccessful, but in every case persisting with unwearied activity, to keep the Indian provisioned, tractable, and in some measure serviceable to his country's need.

Colonel Elliott was reputed a man of tact and wisdom, yet fearless in his courage. It is on record that he tried to dissuade Captain Barclay from dismantling the ramparts, counselled him to await his opportunity to give battle to the American fleet under protection of the guns of Fort Malden, and that he took sides with Tecumseh in opposing the abandonment of the west by General Procter after the disaster which overtook the Canadian fleet on Lake Erie. But there are so many fantastic tales connected with the events of the last month of the western campaign, that it is hard to give credence even to those which have a semblance of truth and wisdom about them. But whatever his views on these two events, he did not forsake the cause of Canada as long as he had physical strength to serve. He was present and took part in Procter's famous retreat, and was one of the fortunate forty of the official staff who made good their escape on the day of battle at Moraviantown. He died the following Spring May 7th, 1814, the direct result of the exposures forced upon a man of his age by the rigorous hardships of warfare, at Ancaster, where he wintered with the British troops, since Amherstburg and the surrounding district was, as a result of the battle of the Thames, in possession of the Americans.

VI.

COMMODORE GEORGE B. HALL
AND LIEUTENANT FREDERIC ROLETTE

THE circumstances which lifted Amherstburg to the mountain peak of its historical dignity in the summer months of 1812 and 1813, were relative to three of the most potent activities connected with the war and its successful prosecution during the fifteen months in which the struggle was continued in the west. Hard by Elliott's Point was the Council House where the Deputy Superintendent of Indian Affairs, Colonel Matthew Elliott, foregathered the Indians to distribute to them their annual presents, supplemented now by equipments of guns, knives and ammunition for war purposes. Here also assembled the stoic Indian warriors, meeting in Council with the white leaders who had induced them to join the British regulars and Canadian militia, to make up the united army which was to defend the domain of Canada from United States aggression. Midway across the stream, on Bois Blanc island, as the whippoor-will began his plaintive evening calls, could be seen the numerous watchfires of the Hurons and kindred tribes, assembled at the call of the fur-traders, and responded to by every member of their families. Standing conspicuous among these was Roundhead, their Chief, while a little farther north could be seen the watchfires of an equally conspicuous warrior, Tecumseh, with his band of Shawnees and other Indian braves, a great gathering of the tribes assembled from far and near, determining the destiny of two nations, their own and that of Canada. In addition, encamped on the commons surrounding the fort, were the yeomen of Essex and Kent, accoutred in the unusual garb of military men, and with them, professional

249

soldiers, a detachment of artillery men and a few hundred of the 41st British Regiment. The presence of all these thousands gave to the place an air of consequence never before or since a part of the community life of the locality.

But Amherstburg was conspicuous and to the forefront in American thought for another reason. It was the naval station of the west. One gun-boat had just been completed, the Lady Prevost, and another was in process of construction to be named the Detroit. In the undertakings which Governor Simcoe had set on foot for the defence of the province, the creation of a marine department found an important place. He had in view the construction of twelve gun-boats for Lake Erie and Captain Alexander Grant was appointed Commodore, to initiate the project. Under his superintendency, a considerable number of merchant vessels had been constructed, but few gun-boats. These were now assembled at Amherstburg, and placed at the service of the army for military purposes during the continuance of the war. His naval fleet, when all were fully equipped, comprised three serviceable vessels, the Queen Charlotte, Lady Prevost, and General Hunter, besides many smaller craft to be used mainly for transportation purposes.

Commodore Grant had given many years of faithful service in developing the potentialities of lake navigation and trade, and had gathered round him a goodly number of efficient seamen, among whom were not a few expert gun-men. When the war cloud loomed up and settled down on the western horizion, he had outlived the allotted span of three score years and ten by fifteen years. Knowing that strenuous requirements would be demanded of the marine department in the conduct of the war, he suggested that his second in command,

George B. Hall, should be promoted to become the head of the department. This was in accord with the policy of Governor Simcoe, who from the first declared his belief that the Provincial Navy should be as far as possible officered and manned by Canadians.

Captain Hall proved himself a capable officer, in his new capacity, and sustained for himself the reputation for efficiency which he held while second in command under Commodore Grant. His fleet, of course, fell far short of the naval equipment visioned by Colonel Simcoe, but such vessels as he had were fully and efficiently manned, by persons who followed a sea-faring life from their youth and thus were inured to the hardships and dangers incident to their vocation. As the Marine Department had of necessity to hold itself in readiness for every emergency during the war, its services were accordingly very varied and in much demand. Its main duties were the transportation of munitions and provisions for army operations, so that the harbour of Amherstburg was a scene of constant activity during the both navigation seasons of 1812 and 1813.

Another very efficient officer appointed about the same time, was Lieutenant Frederic Rolette, a native of Quebec, born there in 1783. He was promoted from a Second Lieutenancy to a First, and given command of the brig, General Hunter, April 25, 1812, in readiness for opening of navigation. Besides fifteen years of service to his credit, in the Provincial Navy when he received this appointment, he had valued experience in the British navy where he received his first training for seamanship. "At an early age," states a well-known author,* concerning this experience, "he enlisted in the British navy,

---

*A. C. Casselman.

and soon had the honor of taking part in two of the greatest naval battles ever fought, and under the most illustrious officer that ever lived. At the battle of the Nile, he received five wounds, and was present at Trafalgar, where the combined naval power of France and Spain was annihilated by Nelson." To this officer is to be given credit for the first successful venture of the war. General Hull had not yet arrived at Detroit, when this first among the many other misadventures overtook his plans. When he arrived at the foot of the Miami rapids, on his way to the Detroit frontier, he placed his sick, the officers' baggage, including some valuable correspondence, and some supplies on the Cayahoga, a packet vessel, with instructions to proceed with them to Detroit. The vessel, proceeding up the river probably unaware of the declaration of war, or the danger of attack, was overtaken by the General Hunter on July third, and captured.

The many-sided activities required of the marine department necessitated that it be manned by persons of resource, and this the Canadian voyageurs of the Upper Lakes undoubtedly were. As an instance, immediately following the withdrawal of General Hull from the Canadian shore, the construction of a battery was commenced by the Canadians opposite Fort Detroit on the left bank of the river. In order to expedite its completion it became the combined work of both the Marine and Engineering departments. While Captain Dixon of the Engineers superintended its construction, Commodore Hall, with his gunboats, provided the material and the munitions, and stood off with these stationed, so as to prevent the workmen from being attacked by any surprise movement of American soldiers. In this way, the work of construction was pushed forward and in readiness to give effective aid when General Brock arrived to give battle to General Hull's army. The batteries finished, their conduct was handed over to the Marine department, as Lieutenant Troughton, the head of the artillery, whose duty it would have been to direct the cannonading, was required by General Brock for another, and what he deemed a very important service, in the plan of attack, the command of the artillery guns on the American side of the river.

"The direction of the batteries was entrusted to Captain Hall and the marine department," reported General Brock, "and I cannot withold my entire approbation of their conduct on this occasion."

So efficient was the service rendered, and so true their aim, that among other damages done to Fort Detroit, a ball was dropped right into the mess room, killing four officers of whom Lieutenant Hanks, lately arrived from Mackinac, was one of them. As has been observed, the men whom Captain Hall had under him were in the main Canadians from the Upper Lakes of whom voyageurs formed the greater number, and they showed themselves equal to the occasion. Had the American army resisted, there is not the least doubt but that these batteries would have given a splendid account of themselves. As it was, there is no gainsaying but that the damage which they effected was an important factor in determining the attitude of mind of General Hull, inducing him to make surrender of the army rather than risk a battle.

No review of the Marine department's activities in that fifteen-month period of the history of the West, could overlook the part they took in the battle of Frenchtown, on the river Raisin, January 22nd, 1813. This battle was the crowning achievement of the Canadian forces under the Generalship of Colonel Henry Procter. The second American army had advanced into Michigan territory and had arrived within twenty-six miles of Detroit. At this place, Colonel Procter had established an outpost in charge of Major Reynolds of the Essex Militia manned by fifty men of that unit and two hundred Indians. On the arrival of the American army Major Reynolds was soon forced to retire, but in his retreat presented a resistance which compelled the enemy to suffer a not inconsiderable loss. Colonel Procter, hastening to his support, gathered together an army of every force within his reach, among whom were the principal men of the Marine department. Their departure from Amherstburg in the early morning of the twentieth, is graphically and succinctly described for us by the young volunteer,* who with his still younger brother, was a part of that impromptu army, and to whom we owe the most reliable extant history of that campaign.

"No sight could be more beautiful than the departure of that little army from Amherstburg. It was the depth of winter; and the river at the point we crossed being four miles in breadth, the deep rumbling noise of the guns prolonging their reverberations like the roar of distant thunder, as they moved along the ice, mingled with the wild cries of the Indians, seemed to threaten some convulsion in nature; while the appearance of the troops winding along the road, now lost behind some cliff of rugged ice, now emerging into view, their polished arms glinting in the sunbeams, gave an air of romantic grandeur to the scene."

The objective of their first day's march was Brownstown, where was located at that time a settlement of Wyandot Indians who were enjoying annual presents from the Canadian

---

*Major John Richardson.

Government, though dwelling then in American territory. Under Roundhead their Chief, these now joined themselves to Procter's army and arrived at the Raisin, the encampment of the American army, and were arranged in order for battle at daybreak on the morning of the twenty second.

In the engagement which immediately took place, the American force became divided into two sections, one part seeking escape in retreat, the other taking shelter in the surrounding houses. The fleeing army became the object of attack by the Indians who fell upon them with such overwhelming and disastrous results, that the engagement has ever since been described as 'the butchery of the River Rasin.' The real battle, however, occurred between the forces comprising the British regulars and Canadian militia and that portion of the American army sheltered within the surrounding houses. These put up a stubborn resistance, which took toll of twenty four killed and one hundred and sixty nine of the Canadian forces wounded, before surrendering. Here the Marine department displayed its usual gallantry, and won for themselves laurels, which even the subsequent disaster of September the tenth, has failed to dim. The gallant Frederic Rolette was four times wounded, Colonel St. George, five times, Captain Irvine and Commodore Hall, once each, and seven others of the officers more or less injured. The Americans taken prisoners, exclusive of those killed by the Indians, exceeded in number more than the whole of the Canadian force altogether, apart from the Indians. It was a gala day in Amherstburg, when the army returned victorious.

But our main interest in the Marine department of Amherstburg, is found in the part it was assigned to play in the conduct of the western campaign in the summer of 1813. In

July of that year, the Americans had completed their naval programme, which was intended to obtain for them naval ascendancy on Lake Erie. During the same month a contingent of forty seamen from the British navy arrived at Amherstburg to reinforce the strength of the marine department, of whom Captain Barclay was the commanding officer. In his apportionment of men to their respective posts, the services of Captain Hall, as Commodore of the fleet, were discontinued. He was instead appointed to the superintendency of the naval docks, a position which carried with it the same remuneration as he was receiving as Commodore. This was an important post, in view of the fact that the Americans had already completed their quota, while the one Canadian boat in process of construction, was still in a half-finished state.

Much criticism has been directed against Captain Barclay for his having discontinued the services of Captain Hall as an officer of the fleet. This has been designated high handed action which was beyond his authority, a criticism which received reinforcement because of the complaint that there was a lack of competent officers even for the few vessels that constituted the Canadian squadron. But doubtless Captain Barclay acted under his best wisdom, and seeing that the construction of the Detroit was the most urgent of the immediate requirements of the Marine department, and that an efficient naval officer, as superintendent of its construction was an important necessity, he did well in selecting Captain Hall for that position. Had this officer suffered a reduction in wages, or if no change in Commodoreship was required, there might be good ground for complaint. Taking, then, the urgency for the completion of the Detroit into consideration, the appointment of Captain Hall to look after this matter,

MALDEN OR AMHERSTBURG, 1800

A town with an historical past, is situated in the township of Malden, Essex County, on the Detroit River. In 1812-15, it was the military outpost of the old Province of Upper Canada, now Ontario. It was laid out in 1796, and early the following year Fort Amherstburg was begun. By some the group of houses outside the fort, to the south, was for a time called Malden, but there does not appear to have been any Fort Malden in the early days. The second fort was known by both names, and the third built after the Rebellion of 1837, bore the name of Fort Malden. This view of Amherstburg is from Elliot's Point, looking up the river Detroit. To the left is Bois Blanc Island, and across from the head of it are the buildings of Fort Amherstburg. The shipyard with a vessel on the stocks may be seen about the middle of the main shoreline and the houses of the town to the right. The water seen is the steamboat channel. In 1812, a block house was also erected in the upper end of the shipyard, and in 1838 there were two put up on the water front. From the 27th of September, 1813, to the 1st of July, 1815, Amherstburg was occupied by American troops, the British having partly destroyed it before their retreat. On the restoration of peace, however, it again came into British possession.

would appear the best arrangement which could have been made under the circumstances. At any rate, the vessel constructed under his superintendency, brought him no loss of prestige. Its need became so urgent that it was brought out of the stocks incomplete, equipped with guns from the ramparts, and in that unfinished state was taken charge of by Captain Barclay, who directed it to render the one service which it was fated to achieve, the defeat of the American flagship, the Lawrence, the only bright spot in the otherwise disastrous defeat of the Canadian squadron on Lake Erie.

The launching of the Detroit ended the career of Amherstburg as a Naval station. The disaster that overtook this vessel and the whole Canadian fleet is recorded in another chapter. Captain Hall continued a resident of the district, but dwelt on his 800-acre farm further up on the banks of the river. After the close of the war, he was elected a member of Parliament for Essex succeeding Colonel Matthew Elliott. Lieutenant Frederic Rolette was second in command of the Lady Prevost in the Lake Erie battle, and was one of those severely wounded and taken prisoners in that engagement. He remained in captivity till the Fall of 1814 At the close of the war, he was presented with a sword by his native city. He died at Quebec on the 17th of March, 1831.

A review of his activities during the fifteen-month period of the campaign, cannot fail to elicit anything other than a gratifying appreciation of the services which he rendered in behalf of Canadian defence. His capture of the Cayahoga and the eleven batteaux laden with provisions, the energy with which he, in association with Commodore Hall and Captain Irvine under Colonel St. George, led their units into the battle at Frenchtown, and their successful assault on the American soldiers sheltered in the French Canadian houses on the battle-field, where he was severely wounded and barely escaped with his life, the courageous support which he gave to his leader as second in command of the Lady Prevost in the naval battle of Lake Erie, all bear testimony not only to his own individual worth, but also to the plucky spirit with which the settlement defended itself in the face of the superior numbers of the invading enemy.

There was one incident connected with the surrender of Detroit, which increases our sympathetic appreciation of the arduous life required of the naval officers during that campaign. The brig, Adams, was one of the prizes of that event. Renamed the 'Detroit', it was used immediately to transport the American soldiers to their various destinations. The irregular soldiers were permitted to return to their homes in Ohio, and were being transported to Buffalo under the terms of agreement of surrender. Lieutenant Rolette was in charge of the Adams, or Detroit, and Captain Irvine of the merchant brig, Caledonia.

"These two vessels," Major John Richardson records, "having reached their destination for landing their prisoners, were then lying, wholly unprotected and unsuspicious of danger, when one dark night they found themselves assailed by two large boats, which had dropped alongside without being perceived until it was too late for anything like effectual resistance. The Detroit was almost immediately carried. . . The surprise of the Detroit and Caledonia was considered by the Americans at that time, a very brilliant feat . . . but it is impossible to look on the exploit in that light." Both vessels being employed in cartel service, it was simply a case of the violation of an honorable treaty agreement.

VII.

ROBERT HERIOT BARCLAY

Captain of the Royal Navy

The Commander of the Amherstburg Fleet, Directing the
Naval Engagement in the Most Glorious Canadian
Defeat During the War of the American Invasion
September 10th, 1813

"MOST Americans, even the well-educated, if asked
which was the most glorious victory of the war,
would point to this battle."

Thus wrote Mr. Theodore Roosevelt, in his book, 'The
Naval War of 1812', concerning the naval engagement which
took place September 10, 1813, on Lake Erie, between a
squadron of six Canadian vessels as the aggressors against, not
an equal number, but nine American vessels. The opinion of
one who subsequently held so high a position in the national
affairs of the United States, and gave during his tenure of the
office of President such pre-eminent satisfaction, cannot be
anything other than highly esteemed. This opinion is
enhanced by reason of his obvious endeavour to be fair in his
treatment of the event, and as he was in a position to rightly
gauge public opinion, there can be no reason for impugning his
judgment.

From the American viewpoint, this achievement, therefore, constitutes the most glorious of all of their undertakings,
not only in the western field, but in every other sphere of their
activities during the whole period of the war, from its
beginning in 1812 until its end in 1815. Since, then, this
engagement figures so prominently in the annals of American

history, it may not be amiss to institute a careful review of all
the circumstances connected with the conflict from a Canadian
viewpoint.

The command of the Canadian squadron was in charge of
a young Scotchman, who came to Amherstburg in the month
of July preceding the engagement, accompanied by forty
others from the British navy destined to serve with him as
officers and sailors on the Lake Erie fleet. Although, at the
time, he was only twenty seven years of age, he had served
under Nelson at Trafalgar, where his heroism won for him
distinction and the loss of one of his arms. Along with about
five hundred others, he was drafted in May, 1813, for the
Canadian service and given a position, when he arrived, with
the Lake Ontario fleet. In July, he was assigned the command
of the Lake Erie fleet, a position of great importance in view
of the fact that, since the surrender of Detroit, the American
nation was putting forth indefatigable efforts to create a naval
fleet, which would be the masters on the waters of Lake Erie,
and without which, they knew, their land armies could not
hope to achieve a successful invasion and the subjugation of
Upper Canada.

On his arrival at Amherstburg, he found awaiting his
command, the nucleus of a fleet—five vessels, comprising one
ship, two brigs, one schooner and one sloop—the Queen
Charlotte, carrying twenty guns; Lady Prevost, twelve guns,
General Hunter, six guns; and two small craft, Little Belt and
Chippeway, carrying, one, a mortar; and, the other, an
eighteen-pounder. These were officered and manned by
Canadians, "provincial sailors, willing and anxious, it is true,
to do their duty, but without that perfection and experience in
their profession, which are so indispensably necessary to the

insurance of success in a combat at sea." *

A second vessel was in construction at Amherstburg, to be called the 'Detroit,' a vessel of good promise if fully equipped with guns and manned by experienced fighting seamen. Unfortunately this vessel was doomed never to realize the promise centred in it. There were two reasons accounting for this miscarriage of Canadian hopes. In the first place, the Americans had their naval programme completed fully one month before the Detroit was ready to leave the stocks. They had achieved the assembly of a fleet of nine vessels, and as there was with them no paucity of resources, their gun equipment showed the same superiority over the Canadian fleet as in the number of their vessels. In addition to so favourable a ratio in respect to their vessels and guns, they were able to man these with the choicest of experienced seamen. Owing to the European war, many frigates lay blockaded in United States ports, and the crews of these became available to supply the needs of their Lake Erie fleet. Over against this American preparedness, there was a second unfavourable circumstance, the lack of experienced officers for the Canadian fleet. Out of the possible four hundred and fifty seamen sent out by Great Britain in May, only forty of these were detailed for service at Amherstburg. The continued call by Captain Barclay for more men fell on the unheeding ears of the official staff above him. At last circumstances could no longer brook delay. The American fleet began to cruise around sometimes in near reach of the guns of Amherstburg. The Detroit had to be taken out of the stocks in an unfinished state, supplied with masts and such other equipment as they could obtain. The great lack was in gun

---

*Roosevelt; Naval War of 1812.

equipment. When the vessel was sufficiently finished to be put into requisition, there was only one thing to do—rob the ramparts—sixes, nines, twelves, eighteens and the two twenty-fours which they had used at the Miami, these were construed into a gun-equipment for a vessel, which, if in a state of proper completion, would be second to none other on the lake even of those of the American fleet.

But strange as it may appear, this vessel, with its motley equipment of guns, was the one which was destined to give the best account of itself in the approaching battle. Selected for the flagship, and under the command of Captain Barclay, had all the others of the fleet been able to encounter the enemy with equal effectiveness, the story of this naval engagement would have been differently chronicled. On the early morning of the ninth, with a favourable breeze inducing them to weigh anchor, this little squadron set out across the lake to make a test of strength of their respective fleets. By daybreak the following day, they had sighted each other, and by noon they were lined up for attack and in action.

The American flagship, the Lawrence, was in charge of the commander of the squadron, O. M. Perry. Captain Barclay singled out this vessel with which he would be specially engaged. For more than two hours, the guns of each sent forth their complements of lead, by which time Captain Barclay was fallen at his post, wounded, while Captain Perry had fled his ship and taken shelter on a safer deck. The first to strike its flag was the American vessel. The two had been tried out on somewhat equal terms, and the Lawrence was worsted.

But how far different was the fate of their respective squadrons. When the roar of the artillery had ended, and the smoke had cleared away, the sea was discovered a holocaust and the decks of the Canadian fleet a shambles. How do we account for so great a disaster in view of the fact that the Lawrence was 'reduced to the condition of a perfect wreck,'* and that, of its crew of one hundred and three persons, eighty three of them, four-fifths of their number, were either killed or wounded, a fate which might have befallen their commander, but for his fortunate escape? The causes leading to this calamity are not hard to discover, and when discovered, they reveal nothing of discredit to either British or Canadian prestige and honour. It is an instance where superior preparedness, and favourable circumstances combined to aid the American, and to place at a proportionate disadvantage, Canadian endeavour. In seamen, in number of their ships, and in weight of metal, the American fleet had decidedly the advantage. This much is conceded by Mr. Theodore Roosevelt, who adds, "with such odds in our favour, it would have been a disgrace to have been beaten."*

This disgrace, however, might have been theirs had it not been for the fickleness of Nature and the vagaries of weather. "The wind which was favourable early in the day, suddenly changed." In these words, Captain Barclay records the fateful circumstance chiefly responsible for their defeat. Committed to fight with sailing vessels, where every one of their movements was made dependent on the wind, they found themselves deserted by the favouring breezes which had brought them into the presence of their enemy. "The weather gave the enemy a prodigious advantage, as it enabled them not only to choose their position, but their distance also, which they did in such a manner as to prevent the carronades of the Queen Charlotte and Lady Prevost from having effect; while the long guns

---

*Theodore Roosevelt: 'Naval War of 1812.'

did great execution, particularly against the Queen Charlotte."

In this paragraph, we have set before us the reason why the Queen Charlotte, a vessel almost equal to the Detroit in staunch construction, and carrying, like it, a complement of twenty guns, did not, with its sister vessel, the Lady Prevost, contribute its expected share towards a successful attack upon the enemy ships. The American vessels were equipped with guns of longer range than any of those on either the Queen Charlotte or Lady Prevost. Before the shifting of the wind, the advantage would have had little effect on the aggregate of results, but when the wind shifted, it enabled the American fleet to choose its position, with the result that they stood off at a distance, close enough to reach these Canadian vessels with their guns, but outside of the reach of the Canadian guns save those of the Detroit. The reader, whose constructive imagination can visualize the helplessness of a sailing vessel, facing a contrary wind, or a calm at sea, will understand why these two vessels did not close in on the Americans and give back in kind as they had received. Can we conceive of any position so provocative of anger or despair —their vessels raked by the metal of the enemy, their comrades falling down all around them, while they were unable to answer back, receiving all but giving none, because an elemental force of Nature had gone over unexpectedly and taken its place on the side and in behalf of the enemy.

A second effect of this change of wind was exemplified in the fate of the Lawrence. When this vessel was compelled, by the superior attack put up by the Detroit, to strike its flag, the Canadian fleet could not, under the circumstance, follow up its triumph by taking possession of the wrecked hulk of their defeated foe. American writers, in consequence of this, have studiously avoided giving the Detroit the credit due to its triumph, some even passing the incident over unnoticed as a matter of inferior consequence. Yet the combat between these two vessels was the one supreme gauge of quality as between their respective seamen. This also is conceded by Mr. Roosevelt.

"Barclay fought the Detroit exceedingly well, her guns being most excellently aimed, though they actually had to be discharged by flashing pistols at the touch-holes, so deficient was the ship's equipment."

Much has been written concerning the failure of the East to come to the support of the West, especially in its supply of men, and both Captain Barclay and General Procter made serious complaint concerning this seemingly unpardonable neglect. But we do not think that in respect to this naval engagement, the disparity of men was accountable for its disastrous issue and the consequent disappointment of Canadian hopes. There was a tendency, then, a very natural one, for the experienced new-comer to over-rate his own qualities, with a corresponding under-rating of those designated 'provincials'.

It is true that the crews of the Canadian fleet, were in the main 'provincials,' a motley aggregation—picked seamen of the British navy and Canadian voyageurs, veteran soldiers of the revolutionary war and school boys of fifteen years of age, white settlers, half-breeds and two Indians—they were all represented in that battle upholding Canada and Canadianism. And did they not, under the circumstances, put up an excellent showing? Had the issue of the battle been dependent alone upon the skill of the respective crews,—their courage, their endurance and their effectiveness—then we are confident that the crews of the six Canadian ships would not have taken a second place to the crews of the American nine in the aggregate of its results. As it was, with all these unfavouring circumstances against them, they entered into the struggle, and within a dozen men of the aggregate—the American casualties were 123 and the Canadian 135—they gave back as they received, man for man, and not one of them, commander or commanded, fled his post save the two Indians. Breton and Canadian, veteran and school boy, experienced and inexperienced, they stood to their guns until they were mowed down by the sheer superiority, not of American manhood, but the range of their guns; the superiority of mechanism backed by an unexpected, but to them, a fortuitous south-east wind. These two forces were placed in their hand, and they used it to destroy these Canadian pioneers, who were seeking to establish for themselves homes in the territory east of the Detroit, and under a form of government whereby they would not be asked to sacrifice their political connection with the Motherland.

Nations may glory, if they wish, in the achievements of mechanical force, but the test of greatness is not in the power, but the use made of it. Mechanism cannot be divorced from morality. Moral law is not confined in its sphere of authority to the relationships existing between individuals, leaving to nations an unrestrained license in the use of destructive powers. The hero of Tom Brown's school days was not the big bully who threw the boot at the little boy on his knees at the side of his bed displaying a courageous fidelity to his upbringing, even although the bully could have "licked the little fellow to a frazzle." Man is greater than mechanism, and courage in man is greater than avarice. The glory of the strong man is not in his strength, but in the use which he makes of it. Judged by the numbers of the Canadians killed, and the number of Canadian vessels wrecked or captured, the battle of Lake Erie lacked nothing in successful American achievement. Judged by a manhood, exhibiting courage, chivalry, endurance, 'even the well-educated' among American citizens may find some difficulty discovering any superiority over that motley defensive force manning the Canadian ships.

Knowing the inferiority of his squadron in both ships and guns, why did Captain Barclay take the aggressive? Why did he sail across the lake in order to fight the enemy in his own waters? Was it the rash act of unreasoned conceit? Or was he forced to it by the whip lash of necessity? This is his own answer.

"The last letter I had the honour of writing to you, dated the 6th instant, I informed you, that unless certain information was received of more seamen being on their way to Amherstburg, I should be obliged to sail with the squadron deplorably manned as it was, to fight the enemy—who blockaded the port—to enable us to get supplies of provisions and stores of every description; so perfectly destitute of provisions was the port, that there was not a day's flour in store, and the crews of the squadron under my command were on half allowance of many things, and when that was done, there was no more. Such were the motives which induced Major-General Procter to concur in the necessity of a battle being risked under the many disadvantages which I labored."

Yet, notwithstanding the compelling motives which induced him to put his six ships up against the American nine, and notwithstanding that his venture had received the concurrence and hearty approbation of General Procter, he

was required, when he was exchanged and sufficiently recovered, to defend his conduct of the battle, before a military
tribunal in accord with the time-honored custom of the Royal
Navy. One would have hoped that this ordeal could have been
spared him in view of what he had already suffered. In the
early weeks of October Major John Richardson, made prisoner
at the battle of Moraviantown, had the privilege of visiting this
young man, as he lay wounded in the cabin of his shattered
vessel, then lying at anchor at Put-in-Bay.

"We found that gallant officer in bed," he records,
"presenting a most helpless picture of mutilation. Pain and
disappointment were on his brow, and the ruddy hue of health
for which he had ever been remarkable, had deserted him. In
short, of his former self there then seemed to be little left
besides his untainted honour."

Shattered thus at his post, surely this was evidence
sufficient of his loyalty to his country and duty. But for the
maintenance of its prestige, naval procedure made it
necessary to discover whether the defeat should be classed
among the honourable or the inglorious ventures of British
naval warfare. After careful scrutiny of every detail, they
found his conduct of the battle, efficient, courageous, and
chivalric, and at no one single point at variance with the
noblest traditions of British seamanship. He fought against
nature and a nation, and he went down under the weightier
forces, physical and mechanical, which had been brought out
against him. After his honourable acquittal, when his services
were no longer required in North America he returned to
Great Britain and took up his residence at Edinburgh. There
he died, May 8th, 1837, aged fifty two years, and was buried
in the historic Greyfriars churchyard of that city.

VIII.

CAPTAIN IRVINE

An Intrepid Upper Lakes Sailor,Whose Brave Services Were Rewarded by His Appointment to the Provincial Navy With Rank of Lieutenant

HERE is yet one other, whose name we should not fail to recall, in making selection of representative Canadians fighting on the western frontier battlefields. This was a Captain Irvine, master of a merchant vessel, but later appointed Lieutenant of the Provincial Navy. When he came into notice first, he was in the service of Angus McIntosh, a fur-trader on the Canadian shore of the Detroit. For a number of years, the residence of Mr. McIntosh was situated at the foot of Moy Avenue, Windsor, overlooking the river. And until very recently though now disappeared, it continued there a landmark of the pioneer days of the district.

The original owner belonged to a prominent family in Scotland, the McIntosh of Moy, who came out to Canada to engage in mercantile business. He succeeded in establishing an important trade among the Indians of the Detroit river district, and also among those dwelling in the Lake Huron region. Though at first a resident of Detroit, with sympathetic sentiments strongly in favor of British connection, he moved his trading house to the Canadian shore, when the cession of Detroit to the American Republic took place. Like many others of the fur-merchants, when the war of the American Invasion took place, he placed his trading vessels at the service of the Provincial Navy, to be used whenever required. Such an occasion arose after the surrender of Detroit, when so many prisoners of war had to be transported to their destinations. It is in connection with this service, that the courage and resource of Captain Irvine was first brought into prominent notice.

It will be sufficient for our purpose to quote, in verbatim, the references to his conduct and worth by Major John Richardson, when writing the story of the various engagements in which he took an honourable part, during the course of the war:

"With this vessel a very gallant affair was connected, only a few days after the capitulation. Agreeable to the terms of this, the irregular forces of General Hull were transported by water to Buffalo, there to be disembarked preparatory to their return into their native state, Ohio, while the regular troops, principally the Fourth United States Infantry, were landed at Fort Erie, with a view to being marched on, as prisoners of war, to Lower Canada.

The armed vessels already named, as having covered our landing on the 16th, were put into requisition for this service, and to these were added the Adams, re-named the Detroit, and the Caledonia, a fine merchant brig, the property of Angus McIntosh, Esquire, of Moy, a few miles above Windsor . . . The Caledonia had her own Captain—Mr. Irvine, a young Scotchman of a peculiarly retiring and amiable disposition and gentlemanly manners, yet endowed with great firmness and resolution of character. These two vessels, having reached their destination for landing their prisoners, were then lying, wholly unprotected and unsuspicious of danger, in the harbour of Fort Erie when, one dark night, they found themselves assailed by two large boats, filled with American sailors and troops which had dropped alongside without being perceived, until it was too late for anything like effectual resistance.

The Detroit was almost immediately carried, but the

young Captain of the Caledonia, which lay a little below her, aroused from his bed by the confusion on board his consort, prepared for a vigorous, although almost entirely personal, resistance. Hastily arming himself, and calling on his little and inexperienced crew—scarcely a dozen men—to do the same, he threw himself in the gangway, and discharged a loaded blunderbuss into the first advancing boat, now dropping from the re-captured Detroit to board the Caledonia. The enemy was staggered, but still they pursued their object, and Mr. Irvine had barely time to discharge a second blunderbuss into the same boat, when he was felled on the deck by a cutlass-stroke from one of the crew of the second party which had boarded him on the opposite gangway. The Caledonia was then secured by her captors, but the Detroit, having grounded, was destroyed.

The intrepidity and self-devotion of Mr. Irvine, whose single arm it appeared, had killed and wounded no less than seven of his assailants, met with that reward it so richly merited. The heads of the Naval Department, anxious to secure so gallant an Officer to the service, tendered to him on his exchange, which took place shortly after, the Commission of a Lieutenant in the Provincial Navy, in which capacity he continued to serve during the whole of the naval operations connected with the Right Division."

Captain Irvine came into notice again in a daring feat successfully achieved in the engagement at Frenchtown:

"The conduct of this latter officer, whose gallantry at the capture of the Caledonia has already been described, was marked on this occasion by the same coolness and resolution. In a forward movement made upon the enemy in the heat of the action, but in which we had been checked by the desperate fire of their riflemen, one of the three-pounders had been abandoned not twenty yards from the fence. The Americans eagerly sought to obtain possession of this, and leaped the breastwork for the purpose of dragging it immediately under cover of their own fire. Their object, however, was seen and frustrated by the British line, which had not retired many yards before it again halted and renewed the contest, compelling the Americans to retire behind their breastworks. Lieutenant Irvine saw the peril of the gun, and, under cover of a heavy fire which was thrown into the enemy at that moment, he advanced, seized the drag rope, and bore it off. This daring feat, performed in the presence of, and between the two armies, was not without its penalty. Mr. Irvine received a wound immediately in the centre of his heel, the ball entering and saturating his boot, which was with some difficulty removed, with blood; and from this effect he suffered for some time. The ball was never found."

Again in the battle of Lake Erie, he had to suffer the irony of fate, and fight against his own former vessel, the Caledonia, now possessed by the Americans and equipped by them with long-ranged guns. In this engagement, he was appointed third command of the Queen Charlotte, a vessel that had with the Lady Prevost, to withstand the carronades of five American vessels, but which could not owing to the sudden change of direction of the wind, draw itself close enough to the enemy vessels to reply in kind. Fighting against such heavy odds and under so unfavourable conditions, the slaughter on the Queen Charlotte was greater probably than on any other vessel. Captain Finnis, the first in command was killed and the second, Lieutenant Stokoe, wounded, both casualties taking place at the very commencement of the

THE HOUSE OF MOY

A pioneer dwelling-house erected by Angus McIntosh, a scion of a noble Scottish family, and one of Detroit's foremost fur-traders, who moved his residence and trading post to the Canadian shore, on the occasion of the cession of that Post to the Americans, 1796. It was built in 1797 by Angus Mackintosh, on farm lot, number 93, first concession, Sandwich, now Windsor.

Windsor was separated from the township of Sandwich in 1854. Mr. Mackintosh, who inherited the estate belonging to the Earldom of Moy, was a factor of the Hudson's Bay Company and resided in this old dwelling until 1830. In 1912 the Moy house was demolished and the property along with the Davis farm nearby, was subdivided in city lots and avenues.

engagement. The command then fell to the lot of Lieutenant Irvine.

"Provincial Lieutenant Irvine, who then had charge of the Queen Charlotte, behaved with great courage, but his experience was much too limited to supply the place of such an officer as Captain Finnis."

Thus reported Captain Barclay concerning his service on that occasion. The same cool intrepidity which characterised his every other activity was displayed here also. The battle ended, he became a prisoner, the lot fated to every member of the six crews who came out of that engagement alive. After suffering for several months the vicissitudes of lot, which prison life among a hostile people affords, he was exchanged in due course, and lived to see the war brought to a close, with the Canadian people continued, a free, independent, and unsubjugated race. To the achievement of that end, he had the satisfaction of knowing, that he had himself contributed a not unworthy part.

One other incident, recorded by Major Richardson, connected with his prison life, it may not be without interest to quote:

"Several gentlemen of the highest respectability in the place were forward in offering attention; and among the first of these was Major Madison. This officer had been himself a prisoner in Lower Canada, from whence he was only recently returned, and impressed with a grateful sense of the treatment he had received, hastened to evince it by various acts of hospitality and courtesy towards ourselves. We became welcome visitors in his family, and frequently accompanied him in excursions to several delightful country seats at some distance from the town.

"As a slight return for the attentions of Major Madison, Lieutenant Irvine of the Navy, had with an ingenuity for which he was remarkable, constructed a vessel in miniature for the daughter of that gentleman. To many of the inhabitants of Kentucky the model even of a frigate in all her parts was a novel sight, and the present was thankfully received. Anxious to tender a similar offering, though in a different quarter, a young midshipman named Campbell, occupying one of the upper rooms, had undertaken a similar task, and devoting himself with all the anxiety and ardor of his years to the completion of the task, soon had the satisfaction of seeing it in a state of great forwardness. Most unfortunately for him, however, he had forgotten that an English flag, even on a bark of those lilliputian dimensions, is ever an offensive image to an American eye; and decked in this fatal oranment, it now lay exposed in one of the windows of his apartment, and was distinctly visible from the street. On the morning of its exhibition, a crowd of persons, delighted at having what they conceived a pretext for insult, rushed in a body up the stairs, uttering imprecations and threats. Having reached the spot where the object of their fury was lying, they seized the luckless ship, and dashed it on the pavement of the street, where it was shattered in a thousand fragments, the leader of the party exclaiming, "You British rascals, if you show your tarnation colors here again, we'll throw you after them." This noble feat being accomplished, they retired, swearing at us all in true Kentucky style, and leaving poor Campbell to brood at leisure over his misfortune."

TECUMSEH
TIPPECANOE, THAMES

**The three Words that sum up the Fateful Tragedy of
the North West Indian Race**

"ENERAL HARRISON: I have with me eight hundred braves. You have an equal number in your hiding place. Come out with them and give me battle. You talked like a brave when you were at Vincennes, and I respected you; but now you hide behind logs and in the earth like a groundhog. Give me answer.—Tecumseh."

In these words we have the legendary challenge of Tecumseh, the representative of the North West Indian nation, calling General Harrison, the representative of the American nation, to a duel of strength. General Harrison was sheltered within the precincts of Fort Meigs, when the Indian chief appeared before that stronghold with his army of eight hundred and challenged him to come out into the open with his eleven hundred, and give them battle. But Tecumseh's eight hundred seemed to General Harrison like unto three thousand. Later, when he had an army under him commanded by nine of the leading military men of the United States, he would come out and fight but not until then.

Tecumseh could not understand why anyone laying claim to bravery, and having an equal chance with the other, should hide himself behind logs and in the earth, 'like a groundhog,' instead of coming out into the open and fighting like a man. But the reason was not far to seek.

Tecumseh, the Indian chief, represented one age; Harrison,

the American General, another. The days of chivalric warfare had ended and the age of mechanism had taken its place. The Indian belonged to an age when man gloried in himself, in his superior personal qualities and his achievements in virtue. The American represented the incoming age, which gloried in the discovery of the powers of nature, and the servitude to which these powers could be exploited for the gratification of man, his aspirations and desires, whether good or evil.

Both men represented the enigma of human life—one human being glorying in the destruction and death of the other. While they both agreed that the killing of one another was a glorious work, they differed in their opinions as to the best method by which this was to be done. The weapon of the Indian was the tomahawk, with the scalping knife wherewith to put the finishing touches on his gory work. The weapon of the whiteman was the rifle, whereby he could reach his victim at a distance, while his own body would be protected by taking shelter 'behind logs and in the earth, like a groundhog'. This latter was the civilized method; the former, the uncivilized.

Tecumseh was not domiciled in Canada when the war of the American Invasion began. His native heath was the Indian territory, the great North West, that triangular region north of the Ohio river, but former Canadian territory, which occupied so prominent a place in the disputed issues arising at the close of the American Revolutionary war. According to one of several traditions, a Cherokee Indian woman, whose husband was a Shawnee warrior, became the mother, at one birth, of three sons.

The birth of twins was a most unusual occurrence in the

history of the Indian race; the birth of triplets an unheard of experience. To the untutored mind of the Indian people, this meant the presage of a great future for those so unusually born. Whether or not there is any truth in regard to the triplet tradition, two, in this family of altogether eight sons, became distinguished far above the usual in the history of individual Indians. Tecumseh became renowned as a great orator, statesman and warrior; his brother became a distinguished prophet of the Great Spirit.

Their home was at the junction, where the Stillwater enters the Great Miami river, a place which Nature had made inspiringly beautiful, but the scene also of the gory history connected with the various expeditions undertaken for the punishment or extermination of the North-West Indian. From earliest childhood, these two heard recounted to them, the injuries which their nation suffered at the hands of the whiteman, and the deeds of valor performed by members of their race in defence of their lives and for the protection of their hunting grounds. Both grew up, therefore, with the picture before their youthful vision of a war-chased race, whose liberty and whose hunting grounds must be recovered. With this vision animating their minds, their youthful hearts were challenged to become the deliverers of their nation.

But while both were inspired by the same vision, yet they grew up differently. The one was given to meditation, the other to activity. Tecumseh grew up a man of action and speech; muscular in limb, erect in his carriage, agile in his movements, courageous in his spirit, his youthful mind animated by the one purpose to deliver his race from the coming doom which threatened it. As a boy when he shot his bow in practise, it was a whiteman in his mental vision, who had done injury to his race, that was the object of his arrow. When he flung his tomahawk with whirling force into some neighbouring tree, to see how straight he could fling it, and how deep sink it into the wood, he was thinking of what he would do to any seeking to wrong his people.

But while Tecumseh grew up, fearless and active, his brother sat down, cast his mind backward, and meditated concerning the ways and works of the Great Spirit. Why did He suffer these wrongs to come upon his race? Why did he not interpose Himself in their behalf?

Thus they both, the one from his meditation, the other in his work and play, prepared themselves for the future part they were to play in the history of their race. Under the inspiration of their vision and purpose, they both harangued the trees, as they thought of their own personal destiny, preparing themselves to be great orators, not of Art, but of Nature, devoting this talent as a last effort to save the independence of the Nation and preserve it from extinction.

Now while these two were dreaming dreams, there was another also, a whiteman, William Henry Harrison, getting ready to climb the political ladder which was to bring him distinction and fame through his power to thwart the aims of these two brothers. This man, a Virginian by birth, and a descendant, it is said, of the great Cromwell in lineage, had set his heart towards denying to the Indian, the liberty and the life which his great ancestor had so strenuously fought for in behalf of the British in England. Instead, he had set his ambition upon building the Eighteenth Council Fire, the eighteenth state of the Union, out of the homelands of Tecumseh and Tecumseh's race.

Thus it was that destiny fated these two to cross swords.

Tecumseh, the warrior, would fight under the inspiration of a valorous spirit. Harrison, the politician, would achieve his political aim through the power of mechanism and the mastering weight of great numbers. The latter was backed by the aspiring self-interest of a nation; the other, supported only by the righteousness of his cause.

Tecumseh came first into prominent notice in 1806. A Council was held in respect to a murder which was committed, a certain Pottawatamie Indian being accused of the crime. The Indians were ordered by the whites to produce this accused person, in order that he might be tried and punished according to the laws of whitemen. This not having been done, this Council was called, and there were present about three hundred chiefs, of whom Tecumseh was one. In the deliberations, everything was proceeding satisfactorily, it being ascertained that the Pottawatamie was not the guilty person, when Tecumseh, springing up, began a declamation which lasted for three hours. Commencing with the arrival of the whites, he recounted the sufferings which the Indians had endured at their hands, up to and including that present time. With fiery eye and moving eloquence, he produced incident after incident from these cumulative examples of wrong-doing, and demanded that it was now high time that these cease.

Under the spell of his burning words, the three hundred warriors threw off the stoic calm with which they were accustomed to listen to the harangues of their chiefs, and excitedly shouted their approval as the speech proceeded. Even the old men, leisurely smoking their pipes, had to forego their age-long custom, and join in uttered approval. That day, Governor Harrison discovered that a great Indian leader had arisen, and that the strong-hand policy of General Wayne would need once more to be put into force, if under Tecumseh's leadership, the Indians should refuse to allow any further encroachments upon their territory, and cease consenting to any further alienation of their lands for white occupation and settlement.

His misgivings were enhanced by information which had come to him at the beginning of the year, that a man belonging to the Shawnees tribe, had set himself up for a prophet. The antagonism between their respective aims was now beginning to be publicly manifested, and must lead to withdrawal or defeat on the part of one of them.

It was five years subsequent to this, when the first clash in arms took place, and the battle of Tippecanoe fought with such fateful results to the Indians. The following June after the Council meeting, where Tecumseh distinguished himself as an orator, the Indian Agent at Fort Wayne informed Harrison, then Governor of Indiana, that the Indians had gone "religiously mad." The next year still more alarming news was sent to him, to the effect that about fifteen hundred of them were assembled from many tribes to hear the prophet but assured him that these had no hostile intentions against the whites. The youthful dreams of the two brothers were beginning to shape themselves into practical undertakings though each of them sought the achievement of their aim according to their different traits of character—different, yet both supplementary to the other.

The method of the prophet, as befitting his meditative mind, was religious. If the Indian race was to avert the doom of extinction, upon which the whiteman had apparently set his aim, then they must call to their help the aid of the Great

Spirit. But to obtain his good-will, they must become obedient children, unlearn all the bad they had been taught by the whites—especially must they refrain from war and the use of intoxicants, the two besetting sins of the Indian people. These doctrines he preached to all the surrounding tribes with an eloquence second only to that of his brother. His renown became widespread, and the number of his followers steadily increased. He made his headquarters on the banks, and at the mouth of the Tippecanoe river, and there he gathered round him the followers whom he had converted to his views.

With the increase of the prophet's followers, the misgivings of Governor Harrison also increased. Taking counsel of his fears, he concluded that this preaching Indian was an impostor, who, through the outward semblance of religion, was cloaking a secret design of hostility against the Whites. These fears and suspicions became doubly augmented when he came to know that the prophet was a brother of Tecumseh, the eloquent chief whose leadership and policies, he increasingly feared, augered no good to the white settlers, and certainly not for the land policy upon which he and his nation had set their hearts. To assure himself that he was right, he had personal interviews with the Prophet, listened to the surmisings of friendly chiefs, and sent his confidential agents to establish, if possible, a case against him.

Throughout it all, the Prophet maintained a consistent attitude, protesting that he had the good only of his own people at heart, and sought no evil against the whites. Still he was no poltroon, and was not afraid to tell the emissaries of Governor Harrison, that, notwithstanding their absence of warlike intentions, it was still true that "the Indians had been cheated out of their lands," and "that no sales were good unless made by all of the tribes," in this supporting the principle set down by his brother to govern the conditions required for the sale of Indian lands to whitemen.

But whether it was because he believed in his own estimate of the Prophet, or whether it was in preparation for the Invasion of Canada, at that time everywhere advocated among the Western States, at anyrate, taking advantage of Tecumseh in the south, he marched his army against the Prophet's headquarters. The Indians stoutly resisted, believing that the Great Spirit would come to their assistance at this critical juncture. Overpowered by the weight of Harrison's numbers, there was effected among them a slaughter, second not even to the achievement along the same line of his predecessor, General Wayne. The slaughter was followed by the burning of their habitations and the destruction of their corn, their winter's supply of food. It was a sorry blow to the cause which the two brothers set out to accomplish. As the Great Spirit failed to come to their aid, it left the remnant of the Indians, as they viewed the dead bodies of their warriors strewn in the melancholy magnitude of their numbers along the banks of the Tippecanoe, without faith or hope, and the Prophet without a following.

The modern reader will ask, from what source did this pagan and untutored redman obtain his religious knowledge and the moral principles which he preached? It requires neither research nor intuition to discover their source. His teachings were the exact counterpart of the Moravian Brethren, the missionaries of which sect were especially active among the Indians of the West at that particular time. He doubtless as a youth attended their services and heard their religious doctrines, and gifted with a good memory, would readily make them his own. Professing, like them, to have received them from the Great Spirit, he began in due time to preach them with increasing effect to the Indian tribes as the way to obtain their freedom from the continuance of the wrongs perpetuated against them by whitemen. That he utterly failed was from no lack of endeavour, but the incoming hordes of Europe were becoming too great in number for him and his brother, with their decreasing followers, to successfully cope against.

But although the Prophet's activities were brought to a close by the battle of Tippecanoe, not so Tecumseh's. His day was not to be so soon and so easily ended. When he came back he found himself without a following save perhaps a half a hundred men, so easily are the Indians discouraged by failure. The matured plan of Tecumseh's was more practical than his brother's, depending upon the inherent exertions of the Indians themselves. Taking a leaf from the experience of the American Nation as did his brother from the Moravian teaching, he advocated a confederacy of all of the tribes of the Indians, an Independent Nation set up in Indian territory north of the Ohio. He laid down three principles which might be called the Constitution of his proposed Confederacy:

1. That all of the scattered tribes of the Indians be joined together as one nation.

2. That Indian lands are the possession of all of the Indian people, and not of one particular tribe only.

3. That no sale of lands can be made by one tribe only, but by all of the tribes conjointly.

To these may be added a fourth, as a corollary,—That the Americans having purchased a section of land from the Miamis, this should not be held as binding, but that it should be returned to the Indian Confederacy.

It was Tecumseh's activities and his successes in gathering support to his undertaking, which Governor Harrison feared more than his brother's religion, hence the battle of Tippecanoe with a view to its overthrow. To permit him to carry out his plan of a confederacy, would be to nullify the personal ambitions of Governor Harrison, and put an insurmountable obstacle in the way to the fulfilment of the American policy which had set out to obtain all of the North West Indian lands for white possession and settlement. The land was too good a land to be left the untouched hunting grounds of an Indian nation.

The aims of these two policies were sharply antagonistic. Tecumseh came out fearlessly and without subterfuge, openly avowing himself committed to the policy of an Indian confederacy, to have and to hold the great North-West as their own. This, he declared, carried with it no intentions of war either against the United States or Great Britain, but to carry on as an independent Indian nation in neighbourly relationships with both of them. But the nation which was prepared to go to war with Great Britain that they might obtain possession of this territory, was not likely to tolerate the policy of making it the homeland of an Indian confederacy. The battle of Tippecanoe was the answer of Governor Harrison and the United States nation to Tecumseh's aim.

It must not be supposed that all of the American people were so callous towards the Indian and his future as to accept without criticism the uncalled-for attack on the Prophet's Settlement. Although it was to the material interest of the white population of the United States that General Harrison should turn his army upon the followers of this preaching Indian,right has sometimes to be considered as well as land. They refused therefore to believe that Governor Harrison was justified in making that attack, nor were they so credulous as to believe that the Indians were the aggressors in the conflict. The friends of General Harrison had therefore to come forward in defence of the Tippecanoe battle. Among those who spent reams of paper to this end, was one, Moses Dawson, who has this to say of Tecumseh, his policy, and the effort of the battle of Tippecanoe towards its overthrow:

"The implicit obedience and request which was paid to Tecumseh, by his followers, was astonishing, and, more than any other circumstance, proved him to be one of those extraordinary geniuses which occasionally arise to produce revolutions, and overturn the established order of things. But for the United States, it is not at all improbable, that this man might have been the founder of a mighty empire not inferior to that of Mexico or Peru. He was deterred by no difficulties. His activity, industry and preserverance, supplied the want of a knowledge of letters. For four years he had been in constant motion; one day he might be seen on the Wabash, and in a short time he would be heard of being on the shores of Erie or of Michigan, or on the banks of the Mississippi; and wherever he went, he made an impression favourable to his purpose.— In fact, to take him all in all, he may be justly said to have been the Bonaparte of the West.

Notwithstanding all this, however, the Governor had strong hopes that the fabric which he and his fanatical brother had reared, and which Tecumseh expected to have finished on this last trip of his, would be demolished, and its foundation rooted up before his return; and the event proved that he was, at least, **nearly,** correct."*

This is candid testimony concerning Governor Harrison's Indian policy, and the expectations which he hoped the battle of Tippecanoe would have realized, but so far from having achieved the final overthrow of Tecumseh's aim by that attack upon the Prophet, he was himself doomed to suffer three humiliating defeats, in two of which Tecumseh was to play an important part, before that event would be chronicled among his successful undertakings.

The armed forces which had been directed against the Indians on the Wabash, were a few months later, turned on the white settlements in the Detroit river district, and for the same purpose, the addition of territory to Republican America. This kinship of experience brought Tecumseh and his followers in sympathetic support on the side of Canada. The number of followers which he could muster was small, but his commanding appearance, his martial abilities, his energetic fearlessness, placed him in the front rank of the great Indian chiefs gathered to the standard of British Canada. The call of the traders— Come over and help us—needed only to be uttered to be answered affirmatively by this chieftain. For fifteen months, he stood by the side of these Canadians and fought with them their successful battles against the invading forces of America. In all this he had but one aim—the preservation of their own territory, and the establishment of his hoped-for Confederacy in the region of the Wabash—which meant, of course, the reopening of the old question of the North West Territory, upon which Great Britain and the United States had already spent

---

*A Historical Narrative of the Civil and Military Services of Major-General William H. Harrison, and a vindication of his Character and Conduct, as a Statesman, a Citizen, and a Soldier by Moses Dawson, Editor of the Cincinnati Advertiser.

thirteen years in discussion, and entered into two treaty agreements to effect a settlement.

The Americans, in choosing to strike their first blow at Canada in the West, anticipated that their success in arms would be so strikingly rapid, as to induce the Indians to join them in battle against Britain; if not, then it would compel them to stand aside in timid neutrality. But although man proposes, there is another who disposes of the fate of Nations as well as of individual men. All these anticipations were shattered by the surrender of Detroit. This success was the signal for the gathering in of Indians from all quarters—men, women and children, until their numbers became a weakness instead of strength, to the Canadian cause.

But one great opportunity for which the fall of Detroit made ample provision, failed to materialize. Moreover, it does not appear to have occurred to any military mind at the time that it ought to have been realized, immediately following Hull's surrender—this, the establishment of the mooted Indian Confederacy on the banks of the Wabash. Had Tecumseh, immediately after the fall of Detroit, struck out independently, and with him the followers which this American defeat had foregathered around him, and had they established the banner of an Independent Indian Nation in the North West, serving notice on both Great Britain and the United States to that effect, the dream of his youth would have become the achievement of his manhood prime.

Moreover, the resisting force against the American Invasion of Canada, would have been trebled in strength, and probably the second invasion of western Canada by the armies of America, would not have been recorded in history. The surrender of Detroit made ample provision for the establishment of two Indian armies, well equipped with the cannons, guns and small arms captured from the defeated army of General Hull. Both of these armies, the one established in the vicinity of Vincennes and the other at Malden and Detroit, would be furnished and fed to capacity. Instead, the Indian aid became an increasingly heavy burden on the Canadian commissary, and the difficulty of keeping this efficient was the chief cause of the disasters befalling the Canadian army in the autumn months of 1813.

What Captain Roberts achieved at Fort Mackinac, that, Tecumseh could have duplicated in the Wabash region, for he had now the requisite number of warriors supplied him. The success of the Canadian arms had brought them to his standard. The five thousand white settlers scattered throughout the adjacent territory would have required the attention of American arms to defend, and while doing this their hands would not have been so free for the invasion of Canada. In addition, the establishment of this confederacy following after the surrender of Detroit, would have reinforced the arguments of the Eastern States against the war, and might indeed under the circumstances have brought about its cessation. Tecumseh, and his Indian followers, not having done this, but instead, leaning upon and imposing an increasingly heavy burden on the Canadian Commissary, rendered a service to the Canadian cause which became a matter of doubtful value.

The organization of this second Indian army would have greatly increased the effectiveness of the Malden forces. Eight hundred or a thousand Indians were all the number which the Canadians could advantageously utilize in conjunction with their own regulars and militia. This number,

JOHN NAUDEE
(Oshawahnah)

A noted war-chief, second in command of the Indians at the Battle of the Thames. He was one of a score of Indian chiefs who distinguished themselves in the support of Canadian defence during the War of the American Invasion, among whom Roundhead, Assiginack, Little Knife, and Big Gun, should be especially mentioned for their splendid support given during the activities in the West, 1812-1813. The Battle of the Thames did not end their activities. The retention of Fort Mackinac and Northern Michigan by the Canadians, during the whole period of the war, and the constant menace maintained to American occupation in Detroit river district by the Indians, served to emphasize the incompleteness of the victories obtained by the fourth American army of the Western Campaign, (the other three having been put out of commission by the Canadian forces), in the late months of 1813. Too much credit cannot be given to the Indians, and the fur-traders who assembled them, and in some instances fed them, for the commendable results obtained.

the Detroit and lake regions could have amply supplied. They were splendid in open battle; effective as separate detachments in harassing the enemy on the march or in retreat; so mobile that they could be used to strike a blow at the Mackinac this week, and at Detroit the next. Lack of roads did not hinder them, and the baggage of the regular soldier was to them a non-essential and an encumbrance. But to obtain the greatest good from their services, they should not be employed in mass formations. They had to be employed in activities which would give freedom of play to their personal qualities of initiative and resource. Numbers tended to render them untractable, insubordinate, with a temptation to revert to the savage state. Separate, or in small detachments, they were initiative, resourceful, capable of a stratagem which made one Indian in certain services equal to three whitemen. An army of one thousand would require at least forty white officers, understanding their language and habits, distributed among them in groups of twenty five each, to obtain from them the most effective service. But given this organization, a thousand of them could easily supply an unhappy existence for an invading army three times their size. In the sons of Colonel Matthew Elliott, Captain Caldwell, Alexander McKee, and others, the Canadians could have mustered a sufficient quota of officers for the effective organization of this army.

The promotion of Tecumseh to the rank of a Brigade officer, gave to him an authority which robbed the Canadian commanding officer of that freedom in the disposition of his troops which a General must fully possess, if he would conduct his campaign effectively. No far-reaching strategic activities could be undertaken, if his plans, he knew, might suffer miscarriage at a critical moment by the insubordination or lack of loyal support of his army. But General Procter, not only labored under the disabilities of this uncertain support but he was forced, especially in the second year of the war, to undertake incursions into the enemy country, which, even if successful, could render no ultimate good to the defensive cause which he supported. The aim of Tecumseh was the establishment of an Indian Confederacy on the Wabash; the aim of Procter, to defend Canada from the ravages of an invading army. While these two aims could be made to work out in unison on certain occasions, there were times when the support of the one would appear the desertion of the other. Such a time occurred when General Procter suggested the withdrawal of the troops from the western frontier in the month of September, 1813. The occasion however would never have occurred, had Tecumseh been as strongly entrenched on the Wabash as Captain Roberts was at the Mackinac, and such a position would have been assuredly his, if he had proceeded there and taken the advantage of the opportunity which the fall of Detroit had placed within his reach.

The ranking of Tecumseh by Mr. Dawson as the "Bonaparte of the West," to this, a review of his activities in the Detroit river district lends no support. Had he been, he would not have missed the opportunity which the fall of Detroit put within his easy reach. In fact, there is no similarity anywhere in the traits of character between these two. Tecumseh, noble in appearance, martial in his carriage, magnanimous in his spirit, and with a lofty purpose inspiring him to achievement, was indeed a great Indian chief; but the marvelous intuition, the boldness of adventure which impelled Bonaparte to promptitude in action, and made him the prodigy of his age, these traits were not above the ordinary in

Tecumseh. Had he been a Napoleon, a fortnight after the fall of Detroit, the establishment of his projected confederacy on the banks of the Wabash, would have been an accomplished reality. This assuredly, Napoleon would have courageously undertaken. This masterstroke of military possibility in the western campaign, he lacked both the intuition to see and the readiness of adventure to put into prompt execution. Tecumseh, as an orator, may have had among Indians, no equal. As a statesman, in his aim and method for its realisation, he was second to no known chief. As a warrior, he was only one among many.

In an age given to war, it is not qualities of manhood that count, but those traits that give success in battle. The cruelty of a Wayne, the craftiness of a Harrison, the intrepidity of a Brock, the boldness of adventure of a Nelson, the intuition of a Napoleon, these are the traits that give success to generalship. But even these will sink into unimportance before the over-mastering weight of metal and mechanism. Tecumseh, we admire him for his noble and magnanimous personality, sympathize with him in his lofty aim to save his race from starvation and extinction, but the number of his contemporary chiefs, his equals in war, could be counted by the score.

The surrender of Detroit was a splendid initial success, but it was only the first incident of the campaign. The destruction of the second American army at Frenchtown was almost equal to it in importance, but Tecumseh was not there and therefore did not share in the honors obtained in that engagement. It is the following up of an achieved advantage which tells the story of a successful ending. This advantage was not seized and the door of opportunity opened out by it not entered.* Both Tecumseh and Procter adopted the attitude of watchful waiting upon the American movements. Not so, Napoleon. The movements of battle would be of his choosing, not the enemy's. With him, the Detroit surrender would have been put into immediate use for further achievements of triumphant victory. Tecumseh was a great chief, but not a Napoleon, and because he was not this opportunity given to him in the prime of his manhood was permitted to pass by unobserved and unanswered.

In the battle of the Thames, we have the closing chapter not only of the Western campaign, but in the life of the brave Tecumseh, the overthrow of the castle of hope of Indian owner-ship and occupancy of the great North West hunting grounds. It would require an impossible credulity to designate it, on the part of the Americans, a great or generous victory. The cream of the Canadian forces at Malden had been either taken prisoners or killed on the preceding tenth of September in the naval engagement on Lake Erie. Of the Canadian militia, the greater number of them were at home gathering in their corn or sowing their winter wheat. The remnant of the regulars were divided up into three sections. One section, two hundred, were sent forward as an escort to the wives and baggage of the official staff, and probably had reached their destination before the engagement took place. The second section, one hundred and seventy five, were deputed to look after the gun-boats, and batteaux bringing the supplies up the river Thames. The third section, consisting of British regulars and a remnant of the Canadian militia, their destiny was being determined by

---

*This was Brock's plan of campaign, but his policy was thwarted by Lord Prevost, Tecumseh had a free hand, and the pity is, that his zeal flagged at that psychological moment.

a divided leadership. Of the Indians, the most of them had concluded that they were not going on an uncertain journey through the woods to Ancaster.

Against this decimal point, nine American Commanding Officers, including two State Governors, brought their combined forces—Generals Harrison, Shelby, Cass, Desha, Henry, Comodore O .M. Perry, Lieutenant-Colonel James Johnson, and Colonel Johnson at the head of 1200 Kentucky mounted riflemen, and lest there should be a repetition of former disasters, these were reinforced by a tenth division, a detachment of American Indians.

In the engagement, General Harrison achieved his aim. The 'fabric' of hope which the two brothers had built up in the Indian heart was broken down, completely and forever. Tippecanoe had silenced the Prophet, but the Thames engagement killed his brother. The ambitions of a Governor and the aspirations of a Nation were now fully achieved. No voice would henceforth be lifted up against the alienation of Indian lands, no strong hand wielded to save the race. Tecumseh was dead.

Amidst all the uncertainties and conflicting statements concerning the battle, two things stand out as indisputably proven facts—the first, the fate of the Indian village of Fairfield; the second, the desecration of the body of the dead Chieftain.

Two miles from the battlefield, and on the same side of the river, was situated the Moravian Mission, established in 1792. To this place an American missionary had brought a following of two hundred and fifty American Deleware Indians, in order that they might escape the doom of extinction threatened by reason of the continued conflict between the

Whitemen and Indians of that region. At the close of the battle, the village of this peaceful Christian community was committed to the flames, their crops destroyed, their people chased into the woods and many of them slain.

As to the desecration of the body of the Chief, the following is the accredited story by a participant in the battle. who, as a man of integrity, could have no motive of either inventing or perpetuating a falsehood, but who had every facility for ascertaining its truthfulness and credibility.

"The merit of having killed Tecumseh, belongs to Colonel Johnson. The merit of having flayed the body of the fallen brave, and made razor strops of his skin, rests with his immediate followers. This, too, has been denied, but denial is vain."

"Several officers of the Forty-First, on being apprized of his fall, went, accompained by some of General Harrison's staff, to visit the spot where Tecumseh lay, and there they identified, for they knew well, in the mangled corpse before them, all that remained of the powerful and intelligent chieftain. Of the pain with which his death was regarded, no stronger evidence can be given than in the fact that there was scarcely an officer of the captured Division, who, as he reposed his head upon the rude log, affording him the only pillow that night, did not wholly lose sight of his own unfortunate position in the more lively emotion produced by the fate of the lamented and noble Indian."*

---

*"Richardson's War of 1812" Casselman edition, Pages 212-213.

X,

GENERAL HENRY PROCTER

The Destroyer of Three American Armies Bearing the Stigma of a Great Reproach Because of the Reverse Occurring at Moraviantown

AMONGST writers of the military events occuring in in the Detroit river district, 1812-1813, there seems to be a unanimous consort of opinion that General Procter did not maintain the prestige of British traditions for bravery and efficiency in his conduct of the war in the west, especially in connection with the events following the Lake Erie naval battle. Was he lacking in bold courage, or was he but the victim of untoward circumstance? Tried by a court-martial in connection with his withdrawal from the western field of conflict, he was found guilty of dereliction of duty, reduced to the ranks for six months without pay in punishment, a judgment which has been approved by an almost universal verdict of succeeding generations of public opinion since. The record of his achievements have thus come down to us with the stigma of a great reproach attached to his name. But is this judgment well-founded, and is it just to the record of the man who bears the stigma?

Concerning this judgment, a leading educationist of Ontario writes, "For his reverse at Moraviantown, he was suspended from rank and pay for six months. In opposition to the general verdict of most historians of this war, I have come to the conclusion that Procter was used disgracefully. No account has been taken of the valuable services he performed. With less than one thousand whites, and a very unreliable Indian following, he destroyed three American armies as large as his own. Reinforcements he asked for, were not sent. His soldiers became stale and dispirited because of neglect from headquarters. The defeat at Moraviantown was the inevitable result of this neglect."

An American writer,* chronicling contemporary opinion, gives an opposite viewpoint:

"Procter's situation at Malden (after Barclay's defeat), made necessary on his part, a prompt retreat, unincumbered with baggage; or a vigorous defence of the post committed to his custody. By adopting the former, he would have saved seven hundred veteran soldiers and a train of artillery, for the future service of his sovereign; by adopting the latter, he would have retained the whole of his Indian allies; given time to the militia of the interior to have come to his aid; had the full advantage of the fortress and its munitions—and a chance, at least, of eventual success, with a certainty of keeping inviolate his own self-respect, and the confidence of his followers. Taking a middle course between these two extremes, he lost the advantages that would have resulted from either. His retreat began too late—was much encumbered with women, children and baggage, and at no time urged with sufficient vigour, or protected with sufficient care. Bridges and roads, ferries and boats, were left behind him, neither destroyed, nor obstructed; and when, at last, he was overtaken and obliged to fight, he gave to his veterans a formation which enabled a corps of four hundred mounted infantry, armed with rifles, hatchets and butcher knives, to win the battle "in a single minute." Conduct like this deserves all the opprobrium and punishment it received, and justly led to General Harrison's conclusion—'that his antagonist had lost his senses'."

*Notices of the War of 1812: John Armstrong.

In his defence before the court-martial, General Procter charged his army with a want of firmness, and to this cause he attributed their defeat at the battle of the Thames. In answer to this, Major Richardson, denying the charge, claimed that General Procter had lost the confidence of his troops, by his management of the campaign all the way through, and so far from any glory due to him for the successes attending the British arms for the preceding fifteen months, every engagement but exhibited his capacity to blunder, of which his movements connected with the retreat were the crowning example.

"On what does General Procter ground his claim to be considered as competent to decide upon the success which ought to attend his military movements," he wrote. "Is it on his dispositions of the river Raisin, where, instead of attacking an unprepared enemy, sword in hand, he absurdly and unaccountably apprised them of their danger, giving them ample opportunity to arm and cripple his own force, in such a manner as to render victory for a period doubtful? Is it on the arrangements at the Miami, where he suffered an important line of batteries to be left without the support of even a single company? Is it upon his attack on Sandusky, where he ordered his men to storm before any breach had been effected, without a fascine or scaling ladder, and with axes so blunt that he might have been suspected of treason in suffering them to be placed in the hands of the unfortunate men who perished while fruitlessly wielding them?"

In every engagement, therefore, according to Major Richardson, the inefficiency of General Procter was exemplified, and victory came, in spite of that inefficiency through the efficiency of the army under him.

Here, then, we have three writers, all making comment, two of them unfavorable, and one favorable to General Procter. The one, in defence, and at variance with the findings of the court-martial, rests his case on the successes which the army under General Procter achieved. In the conduct of the war, he had three brilliant victories, and one notable defeat. Balance the one over against the other three, and you have not got any foundation for your stigma. The second emphasizes what he might have done, while the third, Major John Richardson, a member of the 41st, and an eye-witness of all existing conditions during the year, expresses the sentiments of the army which fought under his generalship.

When Colonel Procter—he was not made General until after the defeat of the second American army at Frenchtown— arrived at Amherstburg in the midsummer month of 1812, he found the garrison there comprised of two hundred and fifty regular soldiers, to whom he added sixty from the 41st Regiment, whom he had brought there with him. In addition to these there were the militia men of Essex and Kent, and a body of about three or four hundred Indians. Pitted against these was an army of twenty three hundred effective men officered by veterans of the American Revolutionary war. To these we must add the American Regulars doing garrison duty at Fort Detroit, and the whole strength of the Michigan militia. It was a nation at war against a settlement, and already the armies of the Nation had over-run all save the Lake Erie settlement in the immediate neighbourhood of Amherstburg. Filibustering detachments of these three army units had gone as far as Moraviantown commandeering whatever of commodities or live stock these struggling pioneers had accumulated.

The establishment of the American encampment on the Canadian shore, isolated Amherstburg, cutting off all communication with the east, save by way of Lake Erie. But, on the other hand, apart from inequality of numbers, Detroit was placed in the same precarious situation. It was a strong fortress, it is true, yet located two hundred miles away from the base of its supplies. This was the one vulnerable feature in the invader's plan of campaign. Against this weak spot, Colonel Procter directed his first effort, although in this he was but carrying out the tactics already initiated by Colonel St. George. A detachment of Hull's army in charge of his outgoing mail, was attacked by a small body of Indians under control of Colonel Matthew Elliott's oldest son, with Tecumseh as their chief. These met with inspiriting success, putting to rout the two hundred which comprised the detachment, captured all of their outgoing official mail, which disclosed important information valuable to the Canadian forces in directing their subsequent movements.

This first encounter, though of miniature importance, was instrumental in creating misgivings in the mind of General Hull as he contemplated what it meant to have bands of Indians under white leadership, standing taut, here and there on the highway which connected him with his homeland and his supplies.

In consequence of this episode, Colonel Miller was despatched with a body of six hundred men, with a view to open out, and keep open, his line of communication. Towards the mouth of the Detroit river, at Maguaga, he came in contact with a portion of the Amherstburg garrison, a body of about one hundred and fifty men, supported by about two hundred and fifty Indians. An engagement took place, after which

Colonel Miller returned to Detroit, and reported that he had worsted and routed and chased through the woods the Canadian forces. Notwithstanding this seeming success, the line was not opened out, but instead bands of Pottawatamies followed and clung to his heels all the way back. In these two preliminary incidents of the war, there was no great military achievement, either of success or defeat, yet it intensified the feeling in the mind of General Hull that he was away out in the wilderness, two hundred miles from the base of his supplies, with fur-traders, voyageurs, Canadian settlers and Indians assembling every day with increasing numbers, composing a force that had already intercepted successfully his mail going out and his supplies coming in, and might continue to do so indefinitely. In this state of mentality, Maguaga, although it was exalted by his fellow countrymen to the rank of a great and victorious battle, brought no thought of quiet content to the General's already disturbed mind. There was required now, only the prompt and intrepid action of General Brock, to compel him to seek safety in the unconditional surrender of Detroit and its surrounding territory.

The outstanding achievement of the western campaign accredited to General Procter, was his assault upon, and his utter destruction of, the second of the American armies sent out to invade western Upper Canada. This army, under command of General Winchester, and on its way to the Detroit river district, had reached the Miami river, and had taken up stationary quarters at Fort Meigs. There they prepared themselves, awaiting only an opportune time to fall upon Amherstburg, and after its reduction, then achieve the crowning ambition of the Americans, the recapture of Fort Detroit. On the arrival of the enemy proximate to him, Colonel Procter

established an outpost at the river Raisin as a first measure
of defence, and manned it with a force of two hundred and fifty
men, comprised of fifty of the Essex militia under command of
one of their number Major Reynolds, of Amherstburg, and with
them a body of two hundred Indians. This outpost was
expected to keep a lookout on the enemy's movements, and
apprise General Procter of any untoward events arising. As
the place where they were stationed, named Frenchtown
because of the French settlers dwelling there, was only
eighteen miles distant from Amherstburg, it will be seen that
the enemy were getting ready to make another test of strength
between them and the Canadian forces.

With the beginning of the New Year (1813), there were
manifest evidences of imminent movements impending. By
January the 18th, an expedition was sent forward by the
Americans under General Lewis and with him a force of eight
hundred men to attack the River Raisin outpost. They soon
succeeded in compelling this small body of volunteers to
evacuate their position, but not before they had taken toll of
the enemy of twelve killed and fifty five wounded, with a loss
to themselves of one militia man killed and two Indians. That
same evening, a messenger was despatched to General Procter
at Amherstburg acquainting him with what had happened,
and with it this further information, that General Lewis had
begun to strengthen himself in their evacuated post and that
General Winchester had quit Fort Meigs and had joined his
forces with those of General Lewis at Frenchtown.

On receipt of this message, Colonel Procter decided on
immediate action. Assembling all his effectives, he mobilized
an army of five hundred whitemen, including the regular
soldiers, the militia men and marines, and supporting these, a

body of five hundred Indians. Lest there should be a tendency to laud Tecumseh, in order to effect a smoke screen in respect to Colonel Procter's generalship, it might be well to record that Tecumseh took no part in this engagement. Among the Indian chiefs, the Wyandot, Roundhead, seems to have been the most prominent. The promptitude and despatch with which Colonel Procter acted may be realized in that before daybreak the second morning after the receipt of the message, his troops were on the march to meet the enemy army, and before daylight on the morning of the 22nd, the engagement had commenced. The American army was taken wholly by surprise as indeed they might well be, but they fought stubbornly and well. The battle became a welter of blood, a massacre of men on both sides.

"I was a witness," wrote Major John Richardson to his uncle, concerning this battle, "of a most barbarous act of inhumanity on the part of the Americans, who had fired upon our poor wounded, helpless soldiers, who were endeavouring to crawl away, on their hands and feet from the scene of action, and were thus tumbled over like so many hogs. However, the deaths of those brave men were avenged by the slaughter of three hundred of the flower of Winchester's army, which had been ordered to turn our flanks, but who, having divided into two parties, were met, driven back, pursued, tomahawked, and scalped by our Indians,—very few escaping—to carry the news of their defeat. The General himself was taken prisoner by the Indians, with his son, aide, and several other officers."

The determination with which the battle was fought is evident from the casualties recorded on both sides. On the Canadian side, there were twenty four killed and one hundred and fifty eight wounded. Of the Americans, over five hundred were taken prisoners, and three hundred killed. In the report of it to Governor Shelby, General Harrison described it as "an event which will overwhelm your mind with grief, and fill your whole state with mourning." "The greater part of Colonel Well's regiment, United States Infantry, and the 1st and 5th regiments, Kentucky Infantry, and Allen's Rifle Regiment, under the immediate orders of General Winchester, have been cut to pieces by the enemy, or taken prisoners."

In the readiness with which Colonel Procter went to the aid of Major Reynolds, in the promptitude with which he gathered together his forces, in the rapidity of his movements, in the effectiveness with which he prosecuted the battle, he showed himself efficient, intrepid, determined, resourceful.

There was no military event which occurred in the western Upper Canada campaign on either side, which can be accounted as surpassing the success of this battle, except, perhaps the forced surrender of Detroit under General Brock. Here the Americans matched strength against strength, and the Canadian forces won out. What General Brock predicted would have taken place at Detroit, if General Hull forced a battle, this occurred at the Raisin. This event alone ought to have merited him an undying place for valorous conduct in the memory of successive generations of Canadians. But his reputation for achieving success is not to be supported alone by these two military events, but a third was, immediately following this, added.

The Canadian forces, having got rid of the first two of the American invading armies, found themselves compelled to face a third, which indeed was following up their second in rapid succession under General Harrison, and had reached within fourteen miles of the Raisin, when the attack on that place was staged. Following the defeat of General Winchester, General Harrison gathered the different units together, and took up his quarters at Fort Meigs on the Miami river, where, in addition to the fortifications already there, he began the construction of such other works as he deemed would make his position impregnable. To attack this position, meant, like the two former successes, an intrepid assault on the enemy army upon his own soil. This, General Procter resolved he would undertake.

It was towards the close of the month of April, before he was ready to get under way. In the meantime, the Indian army had been reinforced to include about fifteen hundred warriors. There was little or no change in the numbers of the regulars and the militia, save that the marines, who had accompanied them to the Raisin were now employed on their vessels, and of the casualties killed in that engagement, there were none to take their place. The Canadian army had no source of reinforcements, save the additions to the Indian army from tribes friendly to them. To the end that this branch would be strengthened, Chief Tecumseh had been devoting himself energetically, and indeed was so employed when the battle of the Raisin took place. The success of that engagement brought them readily to his standard.

The march to Fort Meigs was attended with great difficulty, because of the state of the roads from an unusually wet season. However, they made a safe arrival, and succeeded in establishing two effective batteries on either side of the river. On the first day of May these began to play with well directed aim upon the fortification, a heavy fire being kept up for four successive days without intermission. General

General Winchester   General Hull

These two Generals of the First and Second American Armies, both became prisoners of war; General Hull, in consequence of the surrender of Fort Detroit and his army to General Brock, August 16th, 1812, and General Winchester, captured at the Battle of Frenchtown, which took place, January 22nd, 1813.

What the battle of Saratoga was to the Americans in the Revolutionary war, the surrender of Detroit was to the Canadians, in the War of the American Invasion, 1812-1815. The battle of Frenchtown was next, if not equal, in importance.

When General Harrison, accompanied by nine American Generals with their forces, and supported by a body of Indians, arrived in Canada on the 27th of September 1813, to attack Amherstburg and bring about the re-capture of Detroit, he found both places abandoned by General Procter, who, in the meantime, had withdrawn from the West, with a view to joining his forces with the central army defending the Niagara district, in this way robbing General Harrison of the honor of wresting Detroit from the Canadians by force of arms, a feat upon which the Americans had set their heart.

The two American successes of the Western campaign, the naval battle of Lake Erie, and the killing by Colonel Johnson of Tecumseh at Moraviantown, were almost, if not wholly, neutralized in their effect, by the hardships suffered the following winter by the scarcity of food in the district, and the outbreak of a mild attack of cholera in Detroit, which carried off hundreds of the garrison American soldiers remaining in the Detroit river district. When the war had ended in 1815, their western campaign had secured for the Americans no additional possessions and certainly no prestige.

Harrison sent out a courier with a hurry call for reinforcements. General Green Clay, whose army was already mobilized for the campaign, came hastily forward with fifteen hundred Kentuckians, and supply boats laden with provisions and ammunition. Under instructions from General Harrison, he attacked the battery on the left bank while a sortie from the Fort went out to the one on the right. But although both of these undertakings were successful in the initial stages of the conflict, the batteries being unsupported, and that the men from the Fort succeeded in getting back to their shelter with thirty two of the Canadian forces taken prisoners, a different ending fell to the lot of General Clay and his men. With a small band of the regulars and militia taking the aggressive for the recovery of the batteries, and with the main armies under Major Muir and Tecumseh pressing them hard from other quarters, the Kentuckians scattered. Four hundred of them succeeded in getting within the protection of the Fort, but the rest of them, taking to the woods, were mowed down; not more than one hundred and fifty of them succeeded in making their escape. In the meantime General Harrison asked for a truce under pretext of an exchange of prisoners and during the negotiations, succeeded in bringing his supply boats within the precincts of the Fort and obtained under this ruse their much needed supply of provisions and ammunition. The negotiations ended, the bombardment of the Fort was continued, but no breach being effected, and the Indians becoming restless and scattering, General Procter, ten days later, raised the siege and returned with his troops to Amherstburg. The loss to General Clay's division was not less than six hundred and fifty, while General Harrison lost from amongst his men in the Fort, eighty one killed and one hundred and eighty nine wounded. Of the Canadian forces, apart from whatever losses were sustained by the Indians, their total casualties were slightly over one hundred, fourteen killed, forty seven wounded, and forty one taken prisoners.

This engagement brought to a close the first year of operations by the Canadians in defence of Upper Canada West, and with the exception of the skirmish at Maguaga, success crowned their every effort.

The attack on Fort Meigs, although it did not secure that complement of success which the Canadians had hoped, yet it was a fitting conclusion to their first year's operations. Whatever may be said of the second year, there can be no gainsaying that in the first year of defensive action, the Canadian forces put up a gallant and successful resistance. Had General Procter's services terminated then, his name would have gone down on record as one of the great generals of British history. Unfortunately the second year did not secure him equal results. Some claim, that by inefficient generalship, the Canadians lost in the second year all that which was gained in the first.

But this was impossible. The prestige obtained by the surrender of Detroit, the defeat of three successive American armies, the marching forth of the Canadian forces to meet and fight the enemy on his own soil, these achievements stand forever on the pages of history and cannot be unwritten. It was a plucky defence, and nothing can tarnish the glory of that year's achievements. Less than a full battalion of men, fighting without respite or rest for a whole year, and yet there was no failure to their record, save the skirmish at Maguaga, if so be that this was a failure.

It would be nothing to the discredit of these men who thus

defended the hearths of pioneer Upper Canada, had they failed in every instance where they succeeded. But, having succeeded and having well merited their success, Canadians would be unworthy to name themselves a people, if they failed to remember and acknowledge with appreciation the services and sacrifices of life and limb which these soldiers and settlers rendered, in resisting the invasion of their country. The memory of these successes ought to make us pause before allowing ourselves to attach a stigma to the record of their commander who gave leadership to them in all of these, their first-year engagements.

It was inevitable that defeat should come to the Canadians, with such paucity of resources as fell to their lot, if no support should arrive from Great Britain to aid them. Their full strength, including British regular soldiers, militiamen and voyageurs, never exceeded the status of a full battalion, and every engagement reduced their effective strength to the extent of their killed, at least. There was no source from which substitutes for these could be obtained except from Great Britain. The east needed all of their own available men. Yet only in the Marine department were there additions made, and that only to the extent of forty men, all of whom were taken prisoners or killed in the battle of Lake Erie, two months after their arrival.

The Indians were the only source from which the strength of the resisting army could be kept up, and the number of their available warriors has been greatly exaggerated. But these could not be utilized unless they, together with their women and children, were fed.

While the Canadian settlement had to look forward to decreasing strength, the longer the war continued, the American

nation on the other hand, had resources which were limited only
by the willingness of the people to supply them, a willingness,
which, with their desire for war, suffered no diminution, but
rather increased with the progress of hostilities. As soon,
then, as one army was decimated, another was mobilized to
take its place. It is on record that they had an army of ten
thousand mobilized for the second year's campaign against
western Upper Canada, of which seven thousand were on the
march towards the re-capture of Fort Detroit, and the
reduction of Amherstburg immediately following the defeat of
Captain Barclay's squadron, September 10th, 1813.

Though the soldiers, regulars and militia were few in
number, and worn out with incessant service, they still fought
on hopefully until after the naval battle of Lake Erie. Could
they still, they asked themselves after this disaster, breast the
wave of circumstance, and pitting weakness against strength,
win out because their cause was good? So thought Tecumseh,
the Indian chief, but not so, General Procter. The siege at Fort
Meigs left something to be supplied. The siege of Fort
Stephenson was a failure. General Harrison had adopted the
plan of digging himself in, and the Indians were worthless for
assault purposes. Under the circumstances, General Procter
purposed that he would take one desperate chance more, and
stake all on the fleet. Six against nine, that was the ratio of
probability in their favour. If they won, then another year
of manly struggle; if they lost, then he would fall in with the
counsel of the east, withdraw from the west, and take his
stand against the invader in the Niagara frontier. To this
end, they gathered the most valiant of their men, marines,
soldiers, voyageurs,—they robbed the ramparts of guns for
needed equipment, and the fleet struck out on the morning of

the ninth, intrepid, courageous, determined, but the weather-vane, as unstable as an Indian, was against them. They 'fought to a standstill'* but valor against mechanical power was unavailing, and they went down to defeat, six against nine. That day General Procter lost control of his spirit, and with it the confidence of the men, with whom he was hitherto associated in battle.

Which of you seeing two armies of several thousand each coming against you, doth not sit down and consider whether he is able to meet both of these armies with one of less than one thousand? This was the problem which General Procter was forced now to consider. The one army, at Sandusky, had easy access and unhindered, by reason of the destruction of the Canadian fleet, to the district by way of Lake Erie. The other at the Miami, reinforced by twelve or fifteen hundred mounted Kentucky riflemen, could come up on the right bank of the Detroit, and crossing the river north of the Bois Blanc island, land on Canadian territory on the opposite side of the fort. Having landed, these two armies could close in from these two directions, and surrounding Fort Malden easily effect its complete destruction; and with it, its protecting Canadian army. General Procter sat down, took counsel from these circumstances, and laid his plans to withdraw eastward. He called a council of war before whom he laid his plans, but it was opposition which these gave him and not their support.

The first to oppose withdrawal was Tecumseh, the Indian chief. With that eloquence for which he was noted, he denounced the plan as a desertion of the Indian and the Indian cause, and bade General Procter take example from the departed brave, Sir Isaac Brock, and fight to win, if not, then die on the spot which he was entrusted to protect from foreign

*Theodore Roosevelt, Naval War of 1812.

invasion. In this, the Indian Chief was supported by the traders, and a considerable body of the militia men. Eventually, however, a compromise was made, and Tecumseh agreed to withdraw as far east as the Moraviantown, an Indian village of refugee American Delewares, eighty miles inland.

In agreement with this compromise plan, they began to make plans for their withdrawal from the west. On the twentieth of the month, ten days after the destruction of their fleet, the General despatched the sick, his own family and the families of the other officers, with their baggage, accompanied by an escort of two hundred soldiers, whom he instructed to continue eastward until they arrived at Burlington.* Having dismantled and destroyed the fort at Amherstburg, and burnt down all the public buildings that they might not fall into the hands of the Americans and provide convenient resting places for them when they arrived, they withdrew from there on the twenty fourth, having spent a whole fortnight on these preparations. At Sandwich, they were joined by the Detroit garrison, and after sending supplies and ammunition by batteaux and two gun-boats ahead of them by way of the Thames river, they went forward from there on the twenty seventh, having in the meantime followed the same procedure in respect to Detroit and Sandwich as at Amherstburg. And strange coincidence, on the same day, three miles from Amherstburg on the Lake Erie shore, General Harrison landed his army, and as his manner was, began to erect fortifications for their safety, a wise precaution, in view of the fact that the Indians excelled in surprise attacks.

On the second of October, the main body of the Canadian force arrived at Dolsen's, three miles west of Chatham, and camped there for the night, while the Indians went forward

---

*Tecumseh, by Norman Gurd.

and camped at Chatham. On the same day, the Kentucky riflemen made conjunction with Harrison's troops at Sandwich, and began their march eastward, arriving at the mouth of the Thames that night. The Americans were well led—not by their Generals, though there was a host of them to be sure— but by a Canadian settler, Matthew Dolsen, who knew every cow path in the territory, being one of the first pioneers of the Thames river district. Consequently, the American forces moved with a rapidity which could not have been possible, had they to discover the way for themselves. Moreover, the country was honeycombed with men who were ready to serve that magnificent display of American force, bent on the invasion of Canada, and the destruction of their Indian allies. They were made acquainted with every movement of the Canadian forces, the hindrances which they might expect to meet on the way, the creeks which they would have to cross, and the ford which would enable them to cross the Thames, when they found that those whom they were pursuing were on the north, the opposite side from them, of the river.

At some stage of their retiring movements, it is said that Tecumseh and Procter had agreed that the resisting battle should be fought at Chatham, at the junction of McGregor's creek with the Thames river. Whether this was a preconceived plan made by mutual agreement between these two, or whether it was an unexpected demand from the Indian Chief following one of those explosive, impulsive moods, so characteristic of the race, we have now no way of discovering with certainty. The second in command, the Inspecting Field Officer, Colonel Warburton, knew nothing about it, but, however, he informed Tecumseh that he would support any agreement for which Procter was responsible. Choosing a battlefield, subsequent to the event, Chatham doubtless would have provided the better one of the two.

At some seasons of the year, McGregor's creek is a stream of considerable dimensions. A triangle is formed where the two join, and as the banks are high, a crossing could not be made of either without a bridge. A bridge erected on McGregor's creek, was some distance back from the mouth of the stream, affording, therefore, ample battle ground between there and the river, so that the Canadian forces could be stationed with the bridge at their rear, which would afford them a means of retreat, if the battle went against them. It was therefore an ideal place to strike a blow at the advancing enemy. With the creek and river impeding their progress, the Americans would have had to win or follow the fate of their three preceding armies. As the Canadian boats, with the ammunition and stores, were in advance, yet close enough to supply the army, if in need, everything was in readiness favorable to an encounter, which, if there was any possibility of victory, with such a disparity of forces against them, this was surely the place of opportunity. It was the spot selected twenty years before by Governor Simcoe for a military post and naval station, so suitable for that purpose did the situation appear to him then. By his instructions, a military road was constructed southward towards Lake Erie. Skirting either side of the road was an open wood, which could have supplied cover to meet with advantage the riflemen of Kentucky. Besides, the two gun-boats could have been brought back and played against the Americans from the two sides of the triangle one on McGregor's Creek, the other on the Thames, while the third could have been used to rake them from the rear. Here, then, was a formation provided by nature and by favourable circumstance, which might have wrested from the Americans the advantage which they possessed in numbers. A well-laid out plan of battle field, but it was not carried out. Upon whose shoulders shall the failure be laid?

When Colonel Warburton was at Dolsen's he was only three miles, or an hour's march, from this battle ground. They had therefore three days in which to fortify the place, and set themselves in well-disposed positions to meet the enemy. In this interval, General Procter was at Moraviantown, far in advance of his supplies, and with him Captain Dixon, the only engineer capable of overseeing the fortification of the Chatham site, while his second in command was left without any instructions whatever as to his superior officer's intentions. With the failure to carry out this arrangement with Tecumseh, if such an arrangement had been made, a splendid opportunity to give battle to the enemy, if battle was to be given at all, was lost. If defeat met their effort, and there was little fear of that, they could have retreated south a mile, reformed at Indian Creek, the site of a former Indian village of the Neutrals, and there again give battle to the enemy. If they failed this second time, they could have retreated south on the communication road, and through the woods on the Indian trail, over the ridge, make their way eastward, where they would have been fed and sheltered by friendly settlers all along their path.

Instead of a battle, which if even unsuccessful, would have made the Americans pay dearly for their victory, a further retirement to Moraviantown, was ordered, necessitating the loss of their supply and gun-boats, and the one hundred and seventy five men in their charge, all of them, save a half a dozen who made their escape, taken prisoners. Thus a second

time, was the strength of the Canadians, their regulars and
militia, reduced by about one-third of their number. Little more
than 300 remained to take part in the subsequent engagement.
The issue of the battle when it did take place, resulted as was
to be expected. The Americans easily won the day. The
remnant of the regulars and militia were soon surrounded and
after twelve had been killed and thirty six wounded, the rest of
the three hundred and sixty-seven, which comprised the whole
of the Canadian forces, exclusive of the Indians, was soon
surrounded and most of them taken prisoners. General Procter
and his staff, among whom was Colonel Matthew Elliott, and
with them forty dragoons, made a successful escape. The Indian
force, comprising eight hundred warriors, fought bravely and
in a measure successfully until scattered after the death of
Tecumseh, who fell, fighting to the last for the preservation of
his race and their ancient hunting grounds. The American
army, according to accredited testimony, comprised an army
of not less than six thousand, of whom there were from twelve
to fifteen hundred mounted riflemen.

It is useless to waste time continuing the discussion as to
whether General Procter provided for his army a good
formation at Moraviantown, or a poor one; and whether the
forces under him fought for ten minutes or ten hours. The
battle was lost on the eighteenth of September in the council
chamber at Amherstburg. The proper field of battle
was not at Chatham, nor yet at Moraviantown, but on the north
shore of Lake Erie, at the landing place of General Harrison's
army. Immediately after the naval defeat on Lake Erie, the
fiery cross should have been sent out to every settler and trader
and Indian in the West, and there would have foregathered
there every available militia man, every available Indian, and

the full strength of the regulars, an army, not of course equal to the invading hosts in number, but equal to them in fighting capacity. In counselling retreat, while as yet the whole width of Lake Erie separated him from the enemy army, General Procter stood alone. Men are not inspired to great deeds by a leadership which turns its back to the foe. Military prestige is not obtained by admitting defeat. General Procter has never been forgiven for his desertion of the West at a time when the West needed him most. In this he may have followed the canons of reason and sound judgment, but in military affairs there comes an occasion, when a leader, taking unreason as his guide, stakes his all on a daring venture, and winning out on the strength and courage of his purpose alone, is acclaimed by the world the greatest of heroes. Retreat and surrender are words not found in the vocabulary of successful warfare. In military matters, only battles won, count. Battles are not fought to give the enemy an opportunity to cheer and boast. Reckless daring, apparently, was not a strong trait in General Procter's personality, and since a quality needed in a great General, this was his misfortune, not his fault. A man cannot rise above his gifts, and these are god-given, not man-created.

Had General Procter stood his ground, critisisms such as were later directed against him would have had no influence in diminishing the military prestige which his former victories brought him. There is only one set of circumstances, fighting under which, military men will tolerate defeat and acclaim the subjugated soldier, a hero. It is when, fighting for a just cause, with his face to the foe and his back to the wall, he goes down to death, unconquered in spirit, crushed by the overwhelming forces arrayed against him. Barclay fought though a nation and nature were both against him. Tecumseh fought with a premonition of a doom to his race which he was powerless to avert. Procter beat a retreat, even though a destined fate had decreed that the Canadian people should be a free, independent and unsubjugated nation. Hence the blight upon his military prestige, the stigma of reproach to his name.

Was Procter, then, 'used disgracefully'? Actuated by humane and military reasons, General Hull deemed the surrender of an isolated military post better procedure than the creation of a scene of savage butchery. For this ineptitude for which Mr. Roosevelt claims his age was somewhat responsible, he was sentenced to death. For a similar ineptitude in the prime of his manhood, General Procter was forced to relinquish six months salary, and compelled to live as a private citizen among his fellow countrymen for the same period of time. Considering the spirit of the age, and the hectic atmosphere created by military miscarriages, perhaps after all the Canadian Commander escaped without any great cause for complaint.

INDEX

317

ERRATA

Page 158, read "descriptive" for "description" line 2 of editorial foot-note, and "years 1795 and 1796," line 4, same foot-note.

For 'John B. Askin', pages, 96, 211, 236, 242, read 'John Askin, Jr.'

The reader's attention is also called to the following mis-spelled words: Page 36, increasing, continued; 53, Jesuit; 59, it; 64, Riddell; 68, sovereignty; 69, treked, 75, of; 80, Walpole island; 87, casualties; 88, siege; 109, brunette; 114, one; 131, enunciation, societies; 139, he; 140, negotiations; 146, dwelling; 180, judgment; 201, Colonel; 243, inspiring.

Substitute "undertaking" for 'understanding" page 141, line 29; and "Governor" for "General" page 189, line 7.

CPSIA information can be obtained at www.ICGtesting.com
Printed in the USA
LVOW10s0908160713

342982LV00001B/20/P